What Every Woman Needs to Know about her Gut

What Every Woman Needs to Know about her Gut

Introducing the new FLAT Gut Diet™ for IBS from The Gut Experts

Professor Barbara Ryan, MB, BAO, BCh, MD, MSc, FRCPI, Consultant Gastroenterologist and Clinical Professor of Gastroenterology, Trinity College Dublin

&

Elaine McGowan, M.Sc., B.Sc. (Human Nutrition and Dietetics), Dip. Diet., RD, Clinical Dietitian

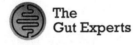

The Gut Experts

sheldon PRESS

First published in Great Britain by Sheldon Press in 2022

An imprint of John Murray Press

A division of Hodder & Stoughton Ltd.

An Hachette UK company

1

A CIP catalogue record for this title is available from the British Library

Trade Paperback ISBN 978 1 529 38826 8

eBook ISBN 978 1 529 38828 2

Typeface by KnowledgeWorks Global Ltd.

Printed and bound in Great Britain by Clays Ltd, Elcograf S.p.A.

John Murray Press policy is to use papers that are natural, renewable and recyclable products and made from wood grown in sustainable forests. The logging and manufacturing processes are expected to conform to the environmental regulations of the country of origin.

John Murray Press
Carmelite House
50 Victoria Embankment
London EC4Y 0DZ

www.sheldonpress.co.uk

To my daughter Meghan, a ray of sunshine.
— Elaine

For Hugo and Owen, and my parents Terry and Tom.
— Barbara

About the authors

Barbara Ryan

Elaine McGowan

Professor Barbara Ryan, MB, BAO, BCh, MD, MSc, FRCPI

Professor Barbara Ryan is a Consultant Gastroenterologist in a Dublin University Hospital and a Clinical Professor of Gastroenterology with Trinity College, Dublin. She studied medicine in Trinity College, Dublin, gaining a coveted Trinity Scholarship and a plethora of awards during her time in university. She completed a doctoral thesis on colorectal cancer and has a Masters in molecular medicine as well as healthcare management. She completed her training in gastro-enterology and internal medicine in Ireland and then spent a number of years working in Germany, The Netherlands and the UK, before returning to Ireland.

Barbara has been working as a consultant gastroenterologist in a university hospital and in private practice for almost 20 years. She has been on a number of advisory boards in Ireland, Europe and the UK. She has a strong research pedigree, with over 100 publications in scientific journals. She has also won a number of research grants to fund ongoing research in a number of areas of gastroenterology including inflammatory bowel disease (Crohn's disease and ulcerative colitis), pancreatic diseases, bone health in patients with digestive disorders and irritable bowel syndrome, to name but a few areas of her expertise.

Barbara has treated tens of thousands of patients over the course of her career and during this time she has gained a real insight into the plight, frustration and poor quality of life of people with many digestive diseases. Given her own background in clinical research, she strongly believes in an evidence-based approach to treating medical conditions, but also believes that this must go hand-in–hand with a holistic approach. The gut and the brain are intimately linked, and one can't be treated without considering the other. She believes that there are not enough experienced medical experts informing and empowering the public regarding their own digestive problems, and that too much information in the public arena is coming from unreliable sources.

Digestive complaints are often embarrassing and people (particularly the older generations) often suffer in silence for years. She would like to help remove the stigma around discussing these important health issues.

Over the past 13 years she has developed a close working relationship with Elaine and they now work together as The Gut Experts. Their mission is the promotion of gut health through education and empowerment of people with digestive disorders, and encouraging eating for health and wellbeing. In 2019, while walking part of the Camino de Santiago, she realized that rather than just complaining about the poor information that patients were being given, she wanted to do something about it! This is her motivation for writing her first book, along with her colleague Elaine, on IBS and the Digestive Health of Women.

Elaine McGowan M.Sc., B.Sc. (Human Nutrition and Dietetics), Dip. Diet., RD, MINDI

Elaine McGowan is one of Ireland's leading private healthcare dietitians and clinical nutritionists. Elaine has been at the forefront of training in dietary solutions for her cohort of IBS patients. She has over 30 years of experience, specializing in diet and gastrointestinal conditions including irritable bowel syndrome (IBS), inflammatory bowel disease (IBD), coeliac disease, diverticular disease and, in more recent years, both small intestinal bacterial overgrowth (SIBO) and functional dyspepsia. Elaine also has a specialist interest in polycystic ovary syndrome (PCOS), underactive thyroid, difficult weight loss and functional gut symptoms associated with these conditions.

In 2010 Elaine travelled to Melbourne, Australia, to meet with one of the main researchers of the low FODMAP diet and was one of the first dietitians in Ireland to train in, and embrace the low FODMAP diet as part of her dietary solutions for IBS. Her dietary philosophy has evolved since her early career and she continually strives to find the most effective dietary solutions for her IBS patients.

Elaine has extensive experience in sports nutrition and has a Masters in sports nutrition: 'A nutritional evaluation of elite Irish athletes'. She worked as a sports nutritionist with the Olympic Council of Ireland for the 1992 and 1996 Olympic Games, and with the sailing and rowing squads for the 2000 Olympic Games. In the past she has worked with the Irish National Rugby Squad, the National Training and Coaching Centre, and notably set up the sports nutrition component of the National Athlete Player Support Programme.

Elaine has provided dietetic consultancy to a wide range of industries. She has designed, piloted, implemented and coordinated workplace wellbeing programmes for several prominent organizations in Ireland.

Like Barbara, Elaine also has a strong interest in the whole person, where she focuses on psychological and emotional wellbeing as part of her holistic approach.

Elaine established her first practice clinic in Ireland in 1992, and since then has successfully used dietary strategies to help over 30,000 patients in her clinics. She also has a wealth of experience working with people to find dietary solutions for optimum health.

Contents

Section 6: FAQs – things we're often asked about

Appendices

Introduction

As a gastroenterologist and a dietitian/clinical nutritionist with a special interest in digestive conditions, we have been working with patients with digestive problems for many years. Between us, we have treated over 60,000 patients with every type of digestive condition, ranging from IBS and functional dyspepsia to severe Crohn's disease, ulcerative colitis, cancers of the digestive system and a range of liver conditions. The overriding concern of all the patients we see, regardless of their condition, is to learn what is wrong with them and how they can get healthy. With this book, we want to give readers the confidence to take control of their gut health and empower them to enjoy life to the fullest.

We find that people get the best results when there is a clear plan or roadmap in place. Uncertainty is undermining and overwhelming. Some conditions can be treated relatively easily. Other conditions, such as Crohn's disease, require quite intensive medical treatment but, thankfully, there have been huge advances in treatment over the past two decades. A condition such as IBS, while not life-threatening, can be life-altering. While there is often no quick fix, which can be disheartening and frustrating for anyone who has been diagnosed with IBS, there have been major advances in our understanding of this condition. We now know that the right, individualized approach for people with IBS can transform their lives and give them back their quality of life.

Some patients who have been diagnosed with IBS have been suffering distressing symptoms for years, soldiering on, often feeling dismissed or unaided by the medical profession. By helping people to correctly identify their main symptoms, the triggers for these symptoms, and the impact of diet and lifestyle, we aim to empower them to make the changes that will get their IBS and other gut problems under control.

The female gut

The digestive systems of men and women are different. This is hardly surprising given how different we are in many other ways – physically, psychologically, emotionally and hormonally. The female body goes through a very complex and extraordinary journey during a woman's lifetime. Like boys, girls undergo significant hormonal changes around puberty. Unlike men, however, a woman's body is then influenced by significant monthly hormonal fluctuations with the menstrual cycle. Any woman who suffers with digestive symptoms, in particular IBS or functional dyspepsia, will know first-hand that the menstrual cycle can have a significant impact on her symptoms. We will explain why this happens and how it can be helped.

Many women become pregnant and give birth, and a woman's body is never quite the same after pregnancy and labour as it was beforehand. This can have knock-on effects on the digestive system.

Women go through menopause, and ultimately enter a post-menopausal phase where the hormones that have played such a huge role all their life, are now at an extremely low level. Many women who have gone through menopause will have noticed a change in their bowel pattern and digestive function around that time. We will explain why this happens and what can be done to help.

The impact of gender identity on gut function and IBS, including genetic males identifying as female and vice versa, is not at all understood at this point, and for that reason is not something that we have discussed in this book. However, it is an important area and one that we will be watching keenly in the future.

Irritable bowel syndrome and functional dyspepsia

Between ten and 11 per cent of the world's adult population has IBS, and 70 per cent of those are women. This equates to about 500 million adults and almost 350 million women worldwide. This is an astounding number of people, and women in particular.

There is another relatively unknown but *extremely* common sister condition of IBS, called functional dyspepsia (FD). Functional dyspepsia affects even more people than IBS (15–20 per cent of adults) AND is more common in people who have IBS. In this book we explain exactly what IBS and functional dyspepsia are, how and why they develop, what investigations should be done, and, in the light of the results, how these conditions can be treated. Most importantly, we describe how treatment must be tailored to the individual.

For some people, the symptoms of IBS or FD are simply an inconvenience and they can manage to live a normal life. Provided that they don't eat certain trigger foods, they are generally fine. For other people, the symptoms can be utterly debilitating and restrictive, preventing them from living a normal life. IBS can affect every aspect of one's life, both personal and professional: it can have an impact on mental and emotional wellbeing, energy levels, sexuality, social interactions and even the ability to exercise.

Many women with these conditions feel that they have been poorly served by traditional Western medicine. They undergo some tests, all of which reveal nothing abnormal. They are given reassurances that there is nothing seriously wrong and that they 'just' have IBS. This simply is not good enough. Women can be embarrassed to discuss their symptoms, and they often feel dismissed and cast adrift to find solutions for themselves. We hope that this book will arm women with the information and confidence they need to advocate for themselves regarding their digestive health, help ensure that they are appropriately (but not exhaustively) investigated and give them evidence-based solutions to help manage their condition.

The interconnectedness of body and mind

Functional disorders such as IBS and functional dyspepsia involve a complex interplay of the body and mind. In our field, we talk about the 'gut–brain axis', which is the intricate and complex two-way flow of information between the brain and the digestive system. Indeed, the gut has been called the 'second brain' because it contains even more serotonin (the 'happy hormone') than the brain itself. Today, researchers have expanded the term to the 'gut–brain–microbiota axis', encompassing the complex role of gut bacteria in this incredible network – a role we explore throughout the book.

Diet

Approximately two-thirds of people with IBS-type symptoms feel that their diet plays a large part. Understandably, this has led many people to try to identify what foods might be causing problems, and to try different exclusion diets in an attempt to control their symptoms. Studies have shown that almost *one third of patients with IBS* have excluded so many things from their diet, that their diet is *nutritionally deficient* (too low in fibre, calcium, iron, vitamin D or vitamin B12) and this can lead to serious health problems in the long term. We have diagnosed early onset osteoporosis developing in a young woman who has excluded dairy from her diet since her early twenties.

A number of dietary regimes, such as the low FODMAP diet, have helped many people with IBS. Sometimes patients can find this a little restrictive and difficult to adhere to in the long term and many patients end up falling back into old habits. For decades, doctors and dietitians have long recognized that certain foods are problematic for patients with IBS. These include fibre, wheat, fructose (fruit sugar), lactose (dairy sugar), the onion and garlic family (known as alliums), sweeteners and a few other items such as caffeine and alcohol.

The role of fibre in gut and overall health and in triggering digestive symptoms has been a little forgotten in recent years, but thankfully fibre is enjoying a renaissance; something we are very happy about. We think of fibre as being a super-food as it is vital for feeding your gut bacteria and maintaining healthy bowel function. In our practices we find that people are often eating far too much or far too little and, like Goldilocks, it is important to get this *just right*!

The Gut Experts' FLAT Gut Diet

We have been using a particular dietary approach in clinical practice for a number of years to help manage problematic symptoms, such as bloating, abdominal pain and irregular bowel habit, all of which are particularly common in IBS and functional dyspepsia. This diet has evolved through our work with tens of thousands of patients, through our combined 50 years' experience. We call this the FLAT Gut Diet, and we are excited to present this to you for the first time in this book.

The FLAT Gut Diet recognizes that a healthy gut needs plenty of variety in the diet. Variety in the diet leads to diversity of the gut bacteria, and this has been shown to be associated with good gut and overall health. More importantly, variety keeps things interesting, and if something is interesting, we are more likely to adhere to it.

The FLAT Gut Diet is based on a Mediterranean-style diet and looks at 'how much' of particular foods we eat, rather than excluding foods altogether. Unless you have a food allergy, most foods do not need to be excluded completely but we do need to look closely at how much, or 'the load' that we are eating.

As we will explain in Section 5 of this book, the FLAT Gut Diet focuses on the most important dietary components that tend to trigger digestive symptoms in patients with IBS. We call these the FLAT Gut Factors.

FLAT Gut Factors

F	Fibre
	Fructans
	Fructose
L	Lactose
A	Alliums
T	Total mind and body health

More than just a diet

We will also explain over the course of this book how diet alone is often not enough to successfully manage IBS and other gut conditions. The 'T' in FLAT stands for 'Total mind and body health'. Given the particularly intimate relationship between the brain and the gut, it is not surprising that things that affect our mental health also have an impact on our gut health. Research has shown how other factors play a crucial role in either controlling or worsening symptoms of many digestive conditions, including IBS and FD. We call these the TEAMS factors and these are:

T	Total mind and body health
E	Exercise
	Regular exercise can help reduce symptoms of IBS
A	Alcohol
	Alcohol can worsen many digestive conditions, including IBS and FD, and intake should be kept to a minimum
M	Mental health
	Mental health needs to be actively nurtured, as stress is often a factor in triggering symptoms
S	Sleep
	It is important to get adequate, quality sleep

We will be asking you to record these factors and to assess their effect on your gut symptoms and overall health and wellbeing. We talk a lot more about the TEAMS factors throughout the book.

For some people, optimizing all these factors may still not be enough to control their symptoms. Certain medications can also have a vital role in managing IBS. While these are never the first-line therapy, they are an important add-on for some people who have particularly problematic symptoms. We look at prescription and over-the-counter medicines as well as natural and herbal remedies, and show how a holistic approach is key to managing IBS and FD.

Our 'mission': to improve quality of life through better gut health

We're both passionate about gut health and digestive conditions. Through our considerable experience, having treated over 60,000 patients between us, we understand from first-hand experience, the frustration and poor quality of life experienced by people with various digestive diseases. We believe strongly in an evidence-based approach to treating medical conditions, but also that this must go hand in hand with a holistic approach.

Our 'mission' is to help you improve your quality of life through better gut health. We will help you do this in a number of ways, which we call 'The Three Es':

1. **Education.** We want you to understand how your gut works, so that you can understand what's going on if you have been diagnosed with a digestive condition. In this book we will be focusing on a number of areas, which are particularly common in women. These are IBS, functional dyspepsia and the effects of female hormones and menopause on gut function.

2. **Empowerment.** By helping you to understand your condition, and removing any uncertainty regarding your diagnosis, you will be empowered to take the correct steps to restore your gut health and equilibrium. Knowledge is power. We want to challenge some of the common myths so that you do not waste your time, energy and money on pursuing treatments and tests that have no proven value.

3. **Eating for health and wellbeing.** By using the FLAT Gut Programme we will help you minimize your symptoms, while eating an interesting, varied and nutritionally balanced diet. But we also want you to understand that sometimes diet alone is not enough and that there are many other things which can help you. You should be confident that your dietary roadmap is the correct one towards better gut health and a better quality of life.

Why are you reading yet another book about your gut?

This might be the first book that you have picked up about digestive problems or perhaps it's the fortieth. But if you are about to read this book, it's probably because all the things you have tried already have not worked for you, or have not worked well enough. If the statements in the box below ring a bell, then this is the book for you.

Fear	You are afraid that there is something serious causing your symptoms that may have been missed or not investigated.

Pain/discomfort	You are experiencing very unpleasant symptoms that are having a negative effect on your quality of life. You want the pain and discomfort to stop.
Hope	You have lost hope or are feeling helpless about your symptoms and wondering if your quality of life will ever improve.
Pleasure	When every food you eat seems to cause you problems, where is the enjoyment? Family meals, eating out and holidays abroad all become something to dread rather than something to look forward to. You need to rediscover some pleasure in life.

How will this book help you?

If you are experiencing frustrating and debilitating gut symptoms, you may want to jump straight to the chapters dealing with dietary solutions in Sections 4 and 5 of this book. We want to empower you to find the solutions to your gut problems, and it will be a lot easier for you and make more sense if you have a good understanding of your gut: how it functions, how your hormones affect it, the importance of the gut–brain axis and ensuring that you have had appropriate investigations. For this reason, we have gone into some detail (hopefully not too much!) about these things in Sections 1, 2, 3 and 6. You may wish to skip around a bit as each chapter is stand-alone.

After reading this book, you should understand how your digestive system works. If you have been diagnosed with a functional digestive condition such as IBS or functional dyspepsia, we want you to have the knowledge to ensure that you have had the correct investigations and, as far as possible, ensure that you have the correct diagnosis. Every patient does not need every investigation, but there are some baseline tests that should be done in all patients before a diagnosis of IBS is made. Your digestive system and symptoms will no longer be a mystery to you, and you'll have a good understanding of the role played by diet, bacteria, emotional factors, mental health, sleep and exercise in triggering and perpetuating these symptoms. Ultimately this will help you to get these symptoms under control so that you can get on with your life.

Remember, managing any digestive condition is a marathon and not a sprint, so you have to be prepared to make changes and to stick with them to reap the benefits. We are confident that this book will be your roadmap to better gut health and wellbeing.

Barbara and Elaine

Myth-busting, facts and fallacies

1 Being told that you have IBS means 'there's nothing wrong'.

Absolutely not! IBS is a very real condition that affects almost one in ten adults. Tests, however, don't show up any abnormalities. We'll explain why later.

2 IBS is 'all in your head'.

This is untrue and insulting to the millions of people worldwide who live with this condition. It can be worsened by stress and other mental health conditions, but is a very real and frustrating condition.

3 IBS is caused by candida overgrowth.

This is not true. Imbalance of candida, along with the trillions of bacteria, may be playing a role in some people, but candida blood tests and candida diets have no role.

4 Why has no-one heard of functional dyspepsia? It must be very rare.

Functional dyspepsia is even more common than IBS, and just like IBS, doesn't show up on standard tests. It is often wrongly labelled or diagnosed as acid reflux or gastritis. We want to spread the word.

5 IBS only causes diarrhoea.

Wrong again. There are different types of IBS and some people have diarrhoea, some have constipation, whereas others alternate between the two.

6 Plant-based diets/so-called 'clean eating' are the best solutions for IBS.

While plant-based foods are very healthy in general, they can be very problematic for people with IBS. We'll tell you why.

7 Gluten-free diets are the solution to IBS

Many people with IBS feel that their symptoms (particularly bloating) improve if they give up wheat. It's not the gluten in these foods that causes bloating, it is instead a question of monitoring 'how much' of these you eat, rather than having to give up these sorts of foods altogether.

8 IBS never goes away.

The symptoms of IBS can come and go, and in women the symptoms often ease up after menopause. Diet, stress and hormonal factors all play a role.

9 The right diet cures IBS.

Unfortunately at this point in time, there is no cure for IBS. BUT, the right, personalized diet can improve your IBS symptoms immensely. Other factors also need to be looked at, which we call TEAMS, and we discuss these more inside.

10 Meat and animal-based foods are bad for you if you have IBS.

Definitely not. Meat, fish and eggs can be eaten as part of a balanced diet, and do not generally cause IBS symptoms. Because of their lactose content, dairy products need to be portion controlled. We will show you how to do this, with our 'lactose points'.

Section 1

UNDERSTANDING YOUR AMAZING GUT

1

How your amazing gut works

The basics

The big picture: the purpose of your gut

1. Your digestive system (commonly referred to nowadays as your 'gut') extends from your mouth to your back passage (anus). The entire purpose of your gut is to safely extract all the nutrients and goodness from the food you put into your mouth and to absorb these nutrients into the bloodstream so that they can be transported around the body to every single cell. This provides the energy for each cell to perform its function. From the neurons (cells) in your brain, to the cells in your womb and ovaries, your hair follicles and even the cells at the base of your toenails – they all get their energy to survive and thrive, from the nutrients in our diet. That is quite amazing.

2. Secondly, we mentioned that your gut has to extract the nutrients *safely*. Safety is extremely important. The food we eat is covered in bacteria and other microorganisms (viruses and other microbes), some of which are potentially harmful and others beneficial. In addition, we are constantly putting different food proteins into our body. In general 'foreign' (as in foreign to our body, not from overseas!) proteins have the potential to trigger immune or allergic reactions (such as nut allergy or hay fever) and so the gut has to know that the food proteins are not harmful, and that it does not need to react to them. This is called 'immune tolerance' and it is very important. Sometimes this process goes awry and certain diseases develop as a result. The gut has many built-in mechanisms to 'defend' itself and the rest of the body from potential invaders.

3. Thirdly your gut has to transform the solid food and liquid you eat and drink (the process of eating and drinking is known as ingestion) into a mulch that can be absorbed. This liquid mulch then needs to be converted back into a solid/semi-solid end product – known as

poop, faeces, stool or bowel motions – in a form that can be expelled in a controlled manner at an appropriate time. Under normal circumstances we get a feeling that we need to go to the toilet, and, if we are in the middle of a meeting, or doing our shopping or playing golf, we can hold onto that bowel motion until a more appropriate moment. But what about the inconvenience caused by a bout of diarrhoea, when you need to run to the bathroom every time you get the urge to go to the toilet? There are many women who experience this constantly (dashing into the undergrowth when out for a walk, sometimes having accidents) and it is a life-altering and socially devastating problem. We will discuss this in a later chapter.

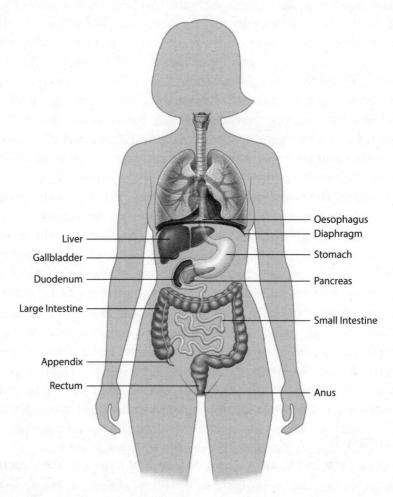

Figure 1 Overview of your digestive system

The anatomy of your digestive system

The mouth and teeth

When we put food into our mouth, the first thing we do is chew the food. This immediately starts the digestive process. Chewing food triggers the release of saliva from the salivary glands just inside our cheeks and below our jaw. Saliva contains amylase, an enzyme that starts to digest carbohydrate and some other digestive products, but also lubricates the food so that we can swallow it. Once the food has been chewed sufficiently, we make a swallowing action that pushes the food into the next part of our digestive system – the oesophagus (or esophagus). Even in the mouth things can go wrong! If you are missing teeth or your muscles are weak (these are particular problems in older people), you may not be able to chew the food properly.

The oesophagus

The oesophagus is a muscular tube about 20 to 25cm long. When you swallow, the food is pushed by your tongue into the top of the oesophagus, where there is a muscle or valve-like structure called the upper oesophageal sphincter (UOS). The UOS relaxes when you swallow to allow the food to pass into the upper oesophagus. As well as controlling the swallow, the UOS also prevents too much air from entering the oesophagus when we are talking and breathing.

Once in the oesophagus, the muscle contractions, along with the effect of gravity if you are sitting upright or standing (hopefully you don't eat lying down!) rapidly propel the food through the oesophagus into the stomach. It only takes a couple of seconds for food (both solid and liquid) to pass through the oesophagus and into the stomach.

At the lower end of the oesophagus, there is another valve-like structure called the lower oesophageal sphincter (LOS). The LOS relaxes after the muscles in the oesophagus contract (pushing the food downwards) and this allows the food to enter the stomach.

The stomach

As you can see in Figure 1, the stomach is a large sac-like structure located just below the diaphragm in the upper abdomen. The stomach can stretch and contract, and it is this stretching that allows us to eat a meal (sometimes a large one) without feeling uncomfortable. The food stays in the stomach for a few hours, and during this time it is churned around like in a cement mixer, where it mixes with the stomach juices and is converted into a porridge-like mulch (think thick, Greek fruit yoghurt). At the lower end of the stomach there is the third muscular sphincter, called the pylorus, and this sphincter opens and closes in a controlled way to allow food to pass in spurts from the stomach into the next part of the gut – the small intestine (small bowel). It takes food and liquid anywhere from half an hour to three hours to empty. The stomach has a number of important functions:

1. **Capacity.** It can stretch and has the capacity to hold the food for a few hours, to let the digestive process begin.
2. **Disinfection.** It produces strong gastric acid (hydrochloric acid), which is an in-built detergent that kills most of the bacteria that we ingest with our food.
3. **Enzymes.** It produces a number of digestive enzymes that continue to digest our food.
4. **Hormones.** The stomach produces a number of very important hormones that play an important role in controlling the release of the semi-digested food into the small intestine. It also produces hormones that are involved in appetite control, or so-called 'satiety'. 'Satiety' refers to the 'satisfied feeling' we get when we have eaten enough. Problems with control of satiety can lead to overeating and obesity or undereating in some conditions.

The small intestine

The 'small' intestine is so-called because it is narrower than the 'large' intestine, but in fact the small intestine is *extremely* long. It measures anywhere from four to six m long (that's about 16 to 22 ft). As you can see in Figure 1, the small intestine is neatly packed into our abdomen but, believe it or not, the surface area, if you flattened it all out, is about the size of a tennis court! Apart from being very long, the small intestine also increases its surface area by having lots of little finger-like projections, called villi (see Figure 2), all the way along its surface. These villi produce a carpet-like surface, and the villi sway like wheat growing in a field so that they are in intimate contact with the contents of the small intestine.

Figure 2 The villi of the small intestine

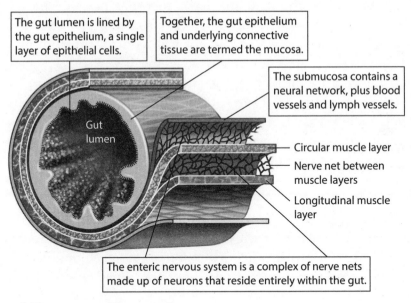

The gut lumen is lined by the gut epithelium, a single layer of epithelial cells.

Together, the gut epithelium and underlying connective tissue are termed the mucosa.

The submucosa contains a neural network, plus blood vessels and lymph vessels.

Gut lumen

Circular muscle layer

Nerve net between muscle layers

Longitudinal muscle layer

The enteric nervous system is a complex of nerve nets made up of neurons that reside entirely within the gut.

Figure 3 The structure of the gut wall

Without getting too technical, it is helpful for you to understand some basic concepts of the structure of the gut wall, in particular the small intestine, as this will allow you to understand how it all works, and later, what happens when something goes wrong. Figure 3 shows a diagram of the basic structure of the gut wall. This basic structure is similar for both the small and large intestine, but there are some differences; in particular, the large intestine does not have villi.

The gut is basically a long cylinder or tube. The space inside the tube is called the lumen, and this is where the food, digestive juices and enzymes all mix together as digestion takes place. In the large intestine there are also trillions of bacteria, the so-called gut microbiota (GM), of which more later. The wall of the gut is made up of a layer of lining cells (a bit similar to the skin on the outside of your body) and these are called the epithelial cells. These cells are in contact with all the contents in the lumen and they are responsible for the absorption of contents from the lumen and secretion (release) of digestive enzymes. Within the wall of the gut, just underneath the epithelial cells, is a rich network of blood vessels, nerve endings and a huge number of immune cells. The blood vessels bring the absorbed nutrients to other parts of the body, as we mentioned earlier, but also bring the immune cells to and from the gut. The nerves in the gut wall control the contractions and movement of the gut as well as being involved in perceiving (sensing) what is going on in the gut.

The outermost layer of the gut is made up of two layers of muscle which, under the control of the nerves in the gut wall, are responsible for the muscle contractions that propel the contents of the gut lumen along from top to bottom.

Functions of the small intestine

Obviously, every part of the digestive system is very important, but the small intestine truly is the powerhouse. You can live without your oesophagus, you can also live without your stomach and your large bowel; however you cannot live without your small bowel – it is vital to survival.

The mulch leaving the stomach enters the top of the small intestine, called the duodenum. It takes anywhere from 30 minutes (very rapid transit) to three to four hours for the contents of the gut to pass along the small bowel (the middle part is called the jejunum) until it reaches the end of the small bowel, the ileum. At the end of the ileum is another sphincter, the ileo-caecal valve, which controls the release of small bowel contents into the large bowel. During these few hours of transit through the small bowel, almost all the nutrients – vitamins, minerals, fats, carbohydrate and protein – are absorbed from the food, and what leaves the small bowel to enter into the large bowel is predominantly waste product, some salts and water. The small bowel can do this amazing job of absorbing all those nutrients because it has such a big surface area (tennis court, remember?), aided by those vital villi.

The main functions of the small bowel are:

1. **Digestion.** It completes the digestion of all ingested food. There are enzymes such as lactase (digests lactose in milk) and many others, in the villi of the small bowel. The pancreas gland empties into the duodenum, and pancreatic juices are rich in a number of digestive enzymes. In addition, the liver produces bile, which is stored in the gallbladder and released into the duodenum, to help with fat digestion.
2. **Absorption.** It absorbs all the nutrients from the diet. This includes all the building blocks we need for a healthy body – including the building blocks of protein, called amino acids, sugars (which are the breakdown products of carbohydrates), fats and vitamins (all the essential vitamins must be obtained from our diet).
3. **Fluid balance.** It is vital in balancing fluid intake and output (along with the kidneys).

The large intestine (the colon)

The colon is wider but shorter than the small intestine. It is approx-imately 1 to 1.5 m (4 to 7 ft) long. The contents of the gut are completely liquid when they pass through the ileo-caecal valve into the colon. One of the main functions of the colon is to absorb the majority of the water from the contents of the gut and to convert something very watery into something solid. This allows us to hold on to the bowel movement until we can access bathroom facilities, which is a lot easier if the contents of the bowel are solid rather than liquid (as I'm sure you're aware after 'that' bout of gastroenteritis when you were on holiday that time...).

We now know that the colon is far more complex than we previously thought, and that the 50–100 trillion bacteria and other microorganisms that make up the GM have far-reaching and complicated functions that exceed those we could ever have imagined even ten years ago. We will be discussing this in the next chapter.

The transit time through the colon is hugely variable and can be anything from eight hours to several days. People with more rapid transit times will tend to pass more frequent bowel motions, whereas someone with slow colonic transit will go to the bathroom less frequently and may have constipation. In some people who are constipated, the colonic transit time can be up to a week!

The rectum. This is the lower part of the colon and sits just above the anus. The rectum is 6–8 in (15–18 cm) long and its main job is to hold the stools inside until such time that we can empty the bowel. When anything (gas or stool) comes into the rectum, nerve endings in the rectum send a message to the brain and the brain then decides if the rectal contents can be released or not. If the brain tells us that it is safe to release the content (i.e. we are in the bathroom), the anal sphincters relax and the rectum contracts, emptying the contents of the bowel. If the contents cannot be released (for example if we're in a shopping mall or out for a walk, far from bathroom facilities), the anal sphincter contracts and the rectum relaxes, and the urge to go to the bathroom temporarily goes away. If we repeatedly override the signal telling us that we need to go to the bathroom, eventually the rectum can become desensitized and this can lead to constipation.

The anus. Also known as the anal sphincter, the anus is the final part of the digestive tract. It is a canal 4–5 cm (2 in) long, made up of the two anal sphincters (internal and external) and a sling of pelvic floor muscles. Nerves in the lining of the upper anus are able to tell if the rectal contents are liquid, gas or solid and hence if it is 'safe' to release the contents or not (it is safe to pass gas when you are out for a walk, but not to pass solid or liquid stools). The sphincter muscles are vital to help control the passage of stool (it is called faecal incontinence when we lose the ability to control the stools). The anal sphincters can be damaged during childbirth, particularly if the labour has been a difficult one, and so too can the muscles of the pelvic floor. These can lead to pelvic floor problems, which we will also discuss in a later chapter.

The functions of the large bowel can be summarized as follows:

1. **Fluid balance.** Absorption of water and salts.
2. **Control.** Control of defaecation.
3. **Gut microbiota (GM).** The colon is home to over 50 trillion bacteria, which make up the GM. The GM is a fascinating and vital part of our digestive system, with far-reaching effects, far beyond our gut (see Chapter 4).
4. **Digestion.** We mentioned that the small intestine is responsible for the digestion of our food. Some foods, such as fibre and certain complex carbohydrates, are not digested by any of the enzymes in the small intestine. These can be digested by the GM and then absorbed from the colon.

When everything works normally, the amazing process of digestion will take place in its entirety without causing any difficulties, but as we will see over the course of this book, things can go wrong at every step of the way, resulting in the various symptoms and conditions that we will discuss.

Gut motility (movement through the digestive system)

There are a number of different kinds of contraction that occur within the digestive system:

1. **Peristaltic waves.** These are contractions that start at some point higher up in the gut and then move along the gut, propelling the contents of the gut lumen along the gut.
2. **Non-peristaltic contractions, or segmental contractions.** These are more unsynchronized contractions that essentially jiggle the gut lumen contents back and forward, ensuring that it is well churned.
3. **Housekeeping waves!** Also known as the migrating motor complex (MMC), these waves occur when we have not eaten for a few hours. This is a strong peristaltic wave that starts at the top of the stomach and sweeps all the way long the small intestine to the colon. It is thought to sweep any particles of undigested food or bacteria along the small bowel and into the colon. These

waves occur approximately six to eight times per day. They only occur when fasting for a few hours and when you are asleep. They explain the typical tummy rumble we have all experienced when we are hungry.

We are not usually consciously aware of this gut motility, but occasionally if peristaltic contractions are stronger than usual, this can cause cramps.

The gut immune system

Seventy per cent of the immune cells in your body are found in your gut. These immune cells are vital to fight off infection/potential invaders, as everything you eat is covered in bacteria and other microbes.[1] As well as preventing potential infection, the immune system also contains regulatory cells that actually dampen down the immune response, and this 'immune tolerance' is very important in helping your gut to learn not to react to 'foreign' proteins and to tolerate the 'good' bacteria in the large intestine.

The immune system in the gut is complex, and good health relies on a delicate balance. If the immune system is overly responsive, you are at risk of developing autoimmune diseases such as Crohn's disease or coeliac disease, but if the immune system is underactive, you may be more prone to infections.

The nervous system of the gut

The gut is controlled by the enteric nervous system, and this system of nerves is incredibly complex and does much more than simply controlling contraction of the gut muscles. Indeed, the enteric nervous system is so complex, that it has been called 'the second brain'.[2] As well as controlling movement of the gut, these nerves are also exquisitely sensitive to what is going on both within the wall of the gut and within the contents of the gut. These nerves sense different things to the 'somatic nerves', which are the nerves that innervate the external tissues such as skin. The gut does not sense pain the same way as we feel pain on the surface of our body and on our skin. If someone were to use a small grasper to take a little 'bite' of your skin, it would be extremely painful, and yet we can take little bite-sized biopsies from all the way along the gut without causing any pain.

The brain registers pain from the digestive system when it contracts (too strongly, causing a cramp) or when it becomes distended and stretched (from a build-up of gas for example). Inflammation in the gut can also activate pain pathways.

We now know that the enteric nerves communicate or 'talk' to the trillions of bacteria within the gut and that this information is then fed back to the brain, via the vagus nerve and other nerve pathways, to create part of what we now call the gut–brain axis.

The gut–brain axis

It is hardly surprising that the brain and the gut are connected, as the brain is connected to every part of our body! So what is so special about the connection between your gut and your brain? It turns out that your gut enjoys a particularly intimate relationship with your brain, compared to the other organs in your body, so much so, that as we just mentioned, the gut has been called your 'second brain' or the 'little brain'.[3] While your brain contains about 100 billion neurones (nerve cells), your gut contains about 500 million neurones, which is pretty impressive, given that your heart has only 40,000 neurones and your kidneys have about one million each. It does seem that there is something special going on in your gut that requires a lot of communication with your brain.

Japanese culture has long recognized the importance of the gut in our bodies, calling it 'onaka' or 'honoured middle' and 'hara' meaning 'centre of spiritual and physical strength'.[2] We now know that there is a constant and complex two-way conversation and flow of information taking place between your gut and your brain. Thus what is happening in your brain (if you are happy, sad, anxious or tired for example) can have a big effect on what is happening in your gut. And, importantly, what is happening in your gut can also affect what is happening in your brain. Think of the 'butterflies in your stomach' you get before an exam, or the need to dash to the bathroom when something nerve-racking is about to happen. All those feelings are due to the gut–brain axis.

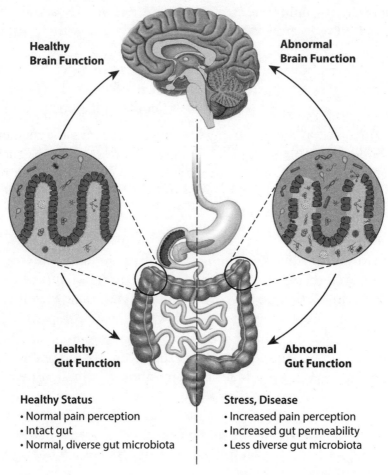

**Healthy
Brain Function**

**Abnormal
Brain Function**

**Healthy
Gut Function**

**Abnormal
Gut Function**

Healthy Status	Stress, Disease
• Normal pain perception	• Increased pain perception
• Intact gut	• Increased gut permeability
• Normal, diverse gut microbiota	• Less diverse gut microbiota

Differences between healthy and impaired gut-brain functioning

Figure 4 The gut–brain axis

What does the gut–brain axis consist of?

As we understand things at this point in time, there are three main communication channels between your brain and your gut: your nervous system, which is the hard-wiring between your gut and your brain, your immune system and your endocrine (hormonal) system. Think of these as being the fast, medium-paced and slower means of communication between the gut and the brain.

The nervous system

The nervous system connects the brain to the enteric nerves, via the vagus nerve and some other nerves that travel via the spinal cord. Information transmitted via nerves travels very quickly and uses chemicals released from the nerve endings, called neurotransmitters. In return, information from emotional and cognitive areas in the brain (involved in the processing of thoughts, feelings and memories) is communicated to the gut. Psychological stress, emotions such as anxiety, fear and anger and physical stimuli such as pain, can bring about changes in the functioning of the gut. These nerve signals can increase gut motility, which can result in diarrhoea, or they can slow emptying of the stomach, which can cause someone to feel nauseated or to vomit.

The immune system

The immune system is ready to act against any potential threat, such as potentially harmful bacteria within the gut; remember 70 per cent of the immune cells in your body are found in your gut. An excess of immune activity within the gut can lead to conditions such as Crohn's disease, ulcerative colitis or coeliac disease, as well as some other less well-known conditions.

The endocrine system

Finally, the endocrine system (hormones produced in glands in other parts of the body) monitors and manages growth, metabolism, as well as motility within the gut. The gut wall has receptors for many different hormones such as cortisol, which is a stress hormone produced by the adrenal gland. It is also rich in oestrogen receptors, which may account for the changes many women see in their gut function over the course of their monthly menstrual cycle. The endocrine system acts more slowly than the nervous and immune systems, but its effects are also longer lasting. This might explain why the presence of stress, which results in higher cortisol levels, may contribute to chronic gut problems, such as IBS.

A two-way conversation

What is happening emotionally and cognitively in the brain affects gut function. But there is also evidence that events within the gut can also affect the pain sensors and pathways in the gut itself, which in turn, relay information back up to the brain and affect the areas of the brain involved in pain sensation. In this way dietary factors or changes in the GM, by altering the local environment within the gut, can lead to increased motility, or distension, which may affect the pain pathways to the brain and reduce the pain threshold.

It is easy to see how this can lead to a vicious cycle:

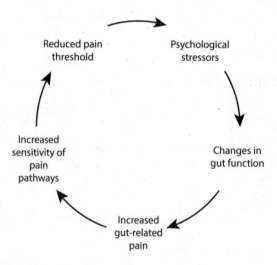

Reduced pain threshold → Psychological stressors → Changes in gut function → Increased gut-related pain → Increased sensitivity of pain pathways → Reduced pain threshold

Take-home messages

- Each part of your gut has a unique function, but your small intestine, with the surface area of a tennis court, is the powerhouse.
- A tummy rumble can mean that your 'housekeeping wave' is kicking in. This is a good thing, so try not to graze continuously.
- Most of the digestive process takes place subconsciously and we become consciously aware of it/develop gut symptoms when something is not functioning correctly.
- The gut and the brain are intimately connected through the gut–brain axis, so a healthy gut is hugely dependent on your mental health.
- The gut microbiota are incredible, and quite simply need a full chapter to talk about them.

2

Organic and functional digestive conditions

The gut can be affected by many conditions. This book will not include an exhaustive list of every known one, but we will briefly introduce you to the concept of an 'organic' versus a 'functional' disorder. This is vital to understanding your gut.

What is an organic condition?

An organic condition, or disorder, is the term used by doctors to describe any medical condition (affecting any part of the body) where observable and measurable abnormalities can be detected. In other words, the results of certain, appropriate tests will show up this condition if you have it. These conditions can be diagnosed through blood or (in the case of gut problems) stool tests. These biological measures are known as biomarkers. Investigations such as x-rays, ultraound CT or MRI scans, upper or lower GI endoscopies can also detect abnormalities in organic conditions.

What is a functional condition?

A functional condition is one that causes symptoms of illness, lack of wellbeing and very often significantly reduced quality of life, yet there are no clear measures by which to make a diagnosis. These conditions can be harder to diagnose definitively and can be very frustrating for those experiencing them, as there is generally no quick 'cure'. The lack of a definitive diagnostic test can generate concern that something may have been missed. People with these conditions also feel that doctors (and family and friends) do not understand the extent to which the symptoms are affecting them and their life, or that their symptoms are somehow imagined.

Thankfully our understanding of many of these conditions has progressed and we know that, while there are often no measurable abnormalities observable in standard investigations, many of these conditions have distinctive characteristics. Psychological factors such a low mood, stress and anxiety play a role in many of these conditions just as they do in every part of our lives, but there are usually other factors or triggers playing a role also.

The two most common functional gut disorders

Irritable bowel syndrome and functional dyspepsia: disorders of gut–brain interaction

Irritable bowel syndrome (IBS) and functional dyspepsia are the most common examples of functional disorders affecting the digestive system. Almost everyone has heard of IBS, but functional dyspepsia is another condition that has not yet achieved the same level of awareness, though it is even more common. Both conditions cause significant problems and reduced quality of life for over 500 million people around the world, and yet, there is no standard diagnostic test with which to make a firm diagnosis. In recognition of the crucial role of the gut-brain axis and connection in these conditions, they are called *disorders of gut-brain interaction* (DGBI).[4]

Up to 50 per cent of women with IBS will also meet the criteria for a diagnosis of functional dyspepsia, and vice versa. Patients are often disappointed when they are told that they have IBS or functional dyspepsia, as they feel that the doctor is telling them 'there is nothing wrong', or is underestimating or failing to recognize the extent of their problem. We want to change that. Yes, the tests are normal, but that does not mean that your gut is functioning normally or that you should feel bad for having symptoms.

Making a diagnosis

If you develop digestive (gut) symptoms, the first step should be to make an appointment to see your GP/family doctor. Based on the symptoms and a physical examination, the doctor will usually consider a number of possible diagnoses to explain your symptoms.

The doctor will then set about doing some tests to check for these possible causes. *If you have a functional condition such as IBS or functional dyspepsia, all the tests will be normal.*

Common organic digestive conditions

There are some relatively common conditions that can affect the digestive system and that need to be considered, or ruled out, before a diagnosis of irritable bowel syndrome or functional dyspepsia is made.

1 Hiatus hernia and gastro-oesophageal reflux disease (GORD/GERD)

This is a very common condition. When a hiatus hernia is present, the top of the stomach slides upwards and 'peeps' above the diaphragm into the chest area. This allows some of the acid produced by the stomach to flow upwards into the oesophagus – 'acid reflux'. Acid reflux irritates the oesophagus, which is not designed to deal with it and causes heartburn, or pain behind the breastbone.

2 Peptic ulcer disease

The word 'peptic' means acid-related. Peptic ulcer disease refers to acid-related irritation or ulcers of the stomach or duodenum. Peptic ulcer disease can have a number of causes, including various medications, particularly aspirin (acetylsalicylic acid) and other anti-inflammatory medications such as Ibuprofen. Excessive alcohol can also play a role. However, one of the most common causes is a bacterium called *Helicobacter pylori* (*Hp* for short). *Hp* can survive in the very acid conditions of the stomach and once a person becomes infected it can persist for decades and can lead to the development of peptic ulcer disease.

3 Coeliac disease

Coeliac (disease) affects people all over the world. About one in 100 Europeans and North Americans are affected, while in Africa, South America and India the incidence is between one in 100 and one in 300 people, indicating that this is one of the most common genetic conditions worldwide. It is almost three times more common in women than in men. Coeliac is caused by allergy to gluten, a protein found in wheat, rye and barley. In patients with coeliac, the lining of the small bowel is damaged by inflammation and this causes damage and eventual shortening of the villi (Chapter 2). Coeliac turns the carpet into flat linoleum (Figure 5), and this affects the ability of the small bowel to do its job. When gluten is strictly removed from the diet, the villi recover again. There is a strong family tendency towards coeliac disease, as this is a genetic condition. If you have a first-degree relative with the condition (a parent, sibling or child) then you have a one in ten chance of developing it yourself.

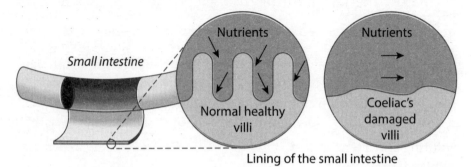

Figure 5 Normal healthy villi vs damaged villi in someone with untreated coeliac disease. Source: Coeliac Society of Ireland

What are the symptoms of coeliac?

Sometimes coeliac causes no symptoms at all and is picked up incidentally, but more commonly it causes some symptoms such as diarrhoea, bloating, abdominal pain or, less commonly, constipation. It can also cause weight loss and vitamin deficiencies (particularly iron, vitamin D and folate), which can result in anaemia and thinning of the bones (osteoporosis or osteopaenia).

How is coeliac diagnosed?

Diagnosis of coeliac is based on two things: a blood test called a tTG (tissue transglutaminase) IgA antibody test and biopsies of the small bowel (taken from the duodenum during an upper GI endoscopy). The blood test is not perfect (about 95 per cent accurate) and biopsies are the definitive test.

How is coeliac treated?

The treatment of coeliac is both very simple and very difficult! It is simple in that all you have to do is avoid gluten (so, a life-long gluten-free diet). This is difficult because gluten is in so many food products nowadays and can be hidden in additives and flavourings. Cross-contamination with crumbs from wheat-containing foods is a big problem particularly when eating in restaurants.

4 Gallstones

Stones can develop in the gallbladder at any time in life. They are much more common in women and as one gets older, but can be found in people of all ages. The gallbladder contracts after you eat, particularly after a fat-containing meal. Sometimes, when the gallbladder contracts, if there are stones present, one may become stuck, leading to an intense pain in the upper central or right part of the abdomen. Typically, this pain comes on within a couple of hours of eating, is very severe (said to be worse than labour pains) and can last for several hours. It is often associated with belching, vomiting and a terrible feeling of upper abdominal bloating. Once the gallbladder relaxes, the pain subsides.

5 Pancreatic problems

Pancreatic conditions are relatively rare but can cause very severe pain, particularly after eating. When the pancreas is damaged and does not produce sufficient digestive enzymes, fats are not digested and absorbed properly. This leads to bowel motions that are oily looking and yellow in colour. If your stools have this appearance you should mention it specifically to your doctor.

6 Small intestinal bacterial overgrowth (SIBO)

Twenty years ago, this was thought to be an uncommon condition. Nowadays with the wider availability of breath tests and the growing awareness of this condition, it is being increasingly diagnosed in people who have digestive symptoms, particularly those with abdominal bloating and excessive wind. We will be discussing SIBO in more detail in Chapter 24.

7 Inflammatory bowel disease

Inflammatory bowel disease (IBD) is a chronic inflammatory condition of the digestive system. Because its abbreviation is similar to IBS, people often get the conditions mixed up. Between one in 500 and one in 1,000 people are affected by IBD and there are two main subtypes: ulcerative colitis (UC) and Crohn's disease. UC causes inflammation in the large bowel (colon) only, whereas Crohn's disease can cause inflammation anywhere from the mouth all the way down to the anus. Despite the fact that Crohn's disease has the potential to affect any part of the digestive system and UC can affect only the colon, both conditions can vary considerably in severity, meaning that someone could have very mild Crohn's disease and be quite well and someone else could have severe UC, and be very unwell.

What causes IBD?

Genetics plays an important role in IBD, and although there is not just one gene involved, most people have a family history of the condition, if not in their immediate family, then in the extended one. Crohn's disease is a little more common in women than in men, whereas UC affects both men and women equally. The peak age for developing IBD is between 15 and 35, although it can occur in younger children and much older adults.

Environmental factors are also very important in the development and activity of IBD.[5,6] Perhaps surprisingly, smoking protects against the development of UC and many patients develop UC for the first time after they have given up smoking. On the other hand, smoking is very bad for Crohn's disease and makes it more active and difficult to treat.

What are the symptoms of IBD?

A primary symptom is diarrhoea with visible blood in the stools. Sometimes people might pass more than 20 bowel motions per day, associated with severe cramps, which is very unpleasant and a serious medical problem. Other symptoms include severe abdominal pains, or an urgent need to go to the toilet with occasional accidents (faecal incontinence), which is obviously very distressing when it occurs.

Diarrhoea that wakes someone from their sleep (nocturnal diarrhoea) is very suggestive of IBD and does not generally occur in IBS.

Not only do IBD and IBS have similar names, but they can also have similar symptoms. However, the following signs are very strongly suggestive of IBD rather than IBS:

- nocturnal diarrhoea
- blood in the stools
- recurrent mouth ulcers (also occur in coeliac disease)
- red eye problems
- low back or joint pain.

How is IBD treated?

There are many different treatments available to treat IBD. Today we have a greater choice of medical treatments than ever before, with new treatments in the pipeline all the time. Treatment depends on the type of disease and the severity of the inflammation. Sometimes surgery is required but not in the majority of cases. Treatment of IBD is a specialized area and is not the focus of this book, but we can say that nowadays treatment of IBD is usually very successful and effective. It is important that any patient diagnosed with IBD should be under the care and supervision of a gastroenterologist.

8 Diverticulosis

Diverticulosis refers to the presence of small out-pouchings of the wall of the colon. They are more common in women than in men, and about 50 per cent of women over the age of 50 will have diverticulosis. Very often they do not cause any problems at all, but can cause lower abdominal pain (particularly in the lower left part of the abdomen), cramps and change in bowel pattern. Very occasionally they can become infected/inflamed and this is called diverticulitis.

9 Cancer

Cancers can develop anywhere in the body, and cancers of the digestive system are relatively common, particularly as we get older. They are very uncommon in young people (less than 40 years of age). Cancers of the digestive system will affect approximately one in 23 women during their lifetime, and mainly when they are older. This means that all the other women (22 out of 23) who have digestive problems do not have, and will not get, cancer. Their problems are caused by other conditions. Many people who develop digestive symptoms are understandably concerned that they might have a cancer. A 20-year-old woman who passes some blood in her stools is much more likely to have haemorrhoids (piles) than cancer. But if a 70-year-old woman starts to pass blood in the stools for the first time, then cancer of the colon would be more likely, although it could also just be haemorrhoids.

Never ignore passing blood in the stool – you should always mention it to your doctor who will assess it and arrange further investigations as necessary.

Warning or 'red-flag' symptoms

Many symptoms can be caused by different conditions, most of them benign, but some not. It is important to talk to your GP/family doctor/gastroenterologist about any new symptoms that you develop and not to dismiss them. Certain symptoms are considered *'red-flag'* or warning symptoms, as they can be associated with cancer. This means that they need to be urgently assessed by a doctor to find the underlying cause.

Warning or 'red-flag' digestive symptoms

- Dysphagia (difficulty swallowing)
- Unintentional/unexplained weight loss
- Early satiety (feeling full after eating a small amount of food)
- Change in bowel habit
- Blood in the stools (either bright red or tar-like stools called melaena)

10 Bile salt diarrhoea/malabsorption

Bile salts/acids are produced by the liver and pass into the small bowel where they help in the digestion and absorption of fats. Any excess of bile salts spills over from the small bowel into the colon, where they can affect the ability of the colon to absorb water, leading to 'bile salt diarrhoea'. Bile salt diarrhoea can mimic IBS-type symptoms and can also worsen diarrhoea in people who have IBS and often goes undiagnosed. Bile salt diarrhoea is more common in people who have had their gallbladder removed. While there are specialized tests available, the simplest way to test for this is to give a trial of a 'bile mop' type of medication, which will rapidly control the symptoms. The effect of these medications on someone's symptoms can be miraculous, and we would almost always give a trial of them to anyone with chronic diarrhoea/loose bowel motions.

If you have digestive symptoms, what do you do next?

If you have developed digestive symptoms, the next step is to discuss them with your GP/family doctor who can help decide on what needs to be done next. There is a lot of information available at our fingertips nowadays, and while Dr Google has helped inform and empower people regarding their own health and wellbeing, it has also led to a lot of anxiety and to conflicting information from unreliable sources.

The next chapter will bring you through how your family doctor/specialist can help diagnose what is causing your digestive symptoms.

Take-home messages

- 'Organic' conditions show up on medical tests.
- 'Functional' gut conditions are now called 'disorders of gut–brain interaction, highlighting the central role of the gut–brain axis.
- Functional conditions do not show up on standard medical investigations, but this does not mean that there is nothing wrong, or that it is 'all in your head'.
- Functional conditions, like IBS and functional dyspepsia are more common than organic conditions and affect 500 million people worldwide.
- You should mention any new gut symptoms to your doctor, do not ignore them.

3

How to investigate gut symptoms

Any investigation of digestive symptoms will depend on what area is affected and, of course, the type of symptoms. We generally divide the digestive system into the upper gastrointestinal (upper GI) tract and the lower gastrointestinal (lower GI) tract. Some people can have symptoms affecting both areas.

The upper GI tract

This includes the oesophagus, stomach and first part of the small bowel (the duodenum). Have a look at Figure 1 again, to remind yourself of the anatomy. It also includes the organs attached to this part of the gut, which are the gallbladder and pancreas. If you have an upper GI problem you might have pain/discomfort/a burning sensation behind the breastbone or just below the breastbone in your upper abdomen. Upper GI symptoms are usually related to either eating or not eating food (fasting). Sometimes the type of food eaten is important, but in some cases the very fact of having eaten something will bring on the symptoms.

Common upper GI symptoms	
Heartburn	A burning feeling behind the breastbone
Belching (burping)	
Dyspepsia	An acid burning/pain feeling in the central upper abdomen (just below the breastbone)
Nausea	A sick feeling/feeling of wanting to vomit
Anorexia	Loss of appetite (not anorexia nervosa)
Post-prandial pain	Pain after eating
Post-prandial fullness	Fullness after eating
Early satiety	Feeling full having eaten a small amount of food
Bloating in upper abdomen	A feeling of being distended
Melaena	Black (tarry) bowel motions, can be caused by a bleeding ulcer

Investigation of upper GI symptoms

Upper GI endoscopy

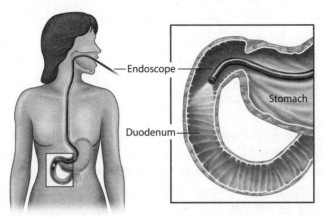

Figure 6 An upper GI endoscopy

An upper GI endoscopy is the cornerstone of the investigation of upper GI complaints. This procedure goes by a number of different names: a gastroscopy, an OGD/EGD or sometimes just an endoscopy. It involves an endoscopist passing a narrow tube that has a camera at the end, into the mouth and down the oesophagus into the stomach and duodenum (Figure 6). This can be done either with or without sedation. Most people opt for some sedation as they may be a little anxious about the tube, but it is down to personal choice and it is very doable without sedation. This procedure allows the clinician to check for any abnormalities and to take samples (biopsies) to check for a number of things, such as coeliac disease or infection with *H. pylori*.

Blood tests

When you present to your GP/family doctor with upper digestive symptoms, they will often check some 'routine' blood tests. These should include a full blood count (to check for anaemia), full biochemical profile and possibly a coeliac antibody test.

Tests for *H. pylori* infection

A test called a urea breath test can be used to diagnose *Hp* infection and it can also be diagnosed using a stool or blood test.

Abdominal ultrasound

An ultrasound is a simple scan of the abdomen and its main organs. You usually have to fast for this test and it involves having a gel-covered probe rubbed around on your abdomen. It is excellent for looking at the organs such as the gallbladder, liver, kidneys and the spleen. Ultrasound would be the first test of choice to check for gallstones.

The lower GI tract

The lower GI tract is significantly longer than the upper GI tract and can be a little more challenging to investigate. Starting from the top down, it includes the small bowel (which can be up to 22 ft long) and the colon (large bowel) that is about 5–7 ft long, resulting in up to 29 ft (over 8 m) of bowel with the potential to cause trouble! A feature of lower GI symptoms is that they tend to be related to passing a bowel motion. Very often they get worse before passing stools and are relieved afterwards, although some people might find that pain gets transiently worse after passing a bowel motion.

Common lower GI Symptoms	
Abdominal bloating	An uncomfortable full feeling, often accompanied by visible distension: 'I look like I'm six months' pregnant'
Flatus	Wind that's expelled from the anus. This can be excessive in some people
Constipation	See below
Diarrhoea	See below
Nocturnal diarrhoea	Waking from sleep with diarrhoea
Borborygmi	Loud gurgling or rumbling sounds coming from the gut
Change in bowel pattern	
Lower abdominal pain/cramps	
Change in appearance of the stools	
Blood or mucus in the stools	
Incontinence	Difficulty controlling the passage of stools causing leakage (accidents)
Obstructive defaecation	Difficulty evacuating the stools or a sense of incompletely emptying the rectum

This is a long list of possible symptoms, and when we talk to patients we find that these terms mean different things to different people, so we'll discuss them in a little more detail. However, first we just want to mention what normal bowel function is.

What is 'normal' bowel function?

The accepted medical description of normal bowel function/habit, is anywhere from three bowel motions per day, to one bowel motion every three days. Most people who do not have any digestive problems have a regular bowel pattern, and generally pass the same number of motions every day and more-or-less at the same time. This is not the case for people who have a digestive problem such as IBS or IBD. In addition to the frequency of bowel motions, normal bowel function usually means that passing bowel motions is painless and easy (as in you should not have to strain excessively etc).

Bloating

This is a sensation described by many patients (particularly those with IBS) and means a sense of fullness of the abdomen. It is often associated with visible distension of the abdomen, which can feel very hard and tense. Bloating is what someone *feels*, and distension is how the abdomen *looks*. Women often say 'I look like I'm six months' pregnant!' The sense of bloating can be eased by passing wind or a bowel motion, but usually not completely. The bloating often goes down overnight and is least marked in the morning, only to build up over the course of the day. We are devoting an entire chapter (Chapter 16) to the how and why of bloating, as this is one of the most distressing and challenging symptoms affecting women with IBS.

Constipation

This tends to mean different things to patients and to doctors. To one patient this can mean passing a bowel motion very infrequently (maybe once a week or once every two weeks), or it can mean passing a hard bowel motion. It can also mean having to strain to pass a bowel motion and a feeling of not completely emptying the bowel. The medical definition of constipation is a little stricter and describes passing fewer than two bowel motions every week or the presence of hard stools that are associated with a difficulty evacuating.

We have seen patients who are complaining of constipation and who are worried because they only open their bowels every two or three days. Firstly this is well within the normal range, and secondly, it is only a problem if it is causing discomfort.

Diarrhoea

Diarrhoea can also mean different things to different people. It includes loose or runny bowel motions, going to the bathroom more frequently than normal, a sense of urgency and, for some, it is also used to describe faecal incontinence. Incontinence or leakage is often an extremely embarrassing problem for people to discuss with their doctor. It is not uncommon in older women who may have some pelvic floor weakness after childbirth, and some women suffer in silence for years because they are too embarrassed to mention it to anyone. Doctors are practised at discussing these things; if you have problems with incontinence, please do not be embarrassed to mention it to your doctor. It is important to remember as we mentioned before, *nocturnal diarrhoea almost never occurs in people with IBS*, so if you are getting up in the middle of the night to open your bowels, you should make an appointment with your doctor to discuss this.

Stool appearance

This can be a difficult topic to discuss with your doctor, but there is nothing to be embarrassed about. The ability to describe the bowel motions has been greatly helped by the Bristol stool chart (and, thanks to the smart phone, we have also been shown real life photos on many occasions!). This stool chart describes stools on a range from 1 to 7. As you can see types 1 and 2 are harder and types 6 and 7 are loose/soft. Normal stools are somewhere in the middle, although many 'normal' women describe stools in the 1 to 3 range, while men tend to have looser stools ranging from 5 to 6.

Stool colour can also vary and in general, looser stools tend to have a lighter colour than harder stools. Very black (tar-like) stools can be a sign of bleeding and also occur if you take iron supplements. This should be mentioned to your doctor. Very pale or yellow-coloured stools can occur in a number of conditions and should also be specifically mentioned to your doctor.

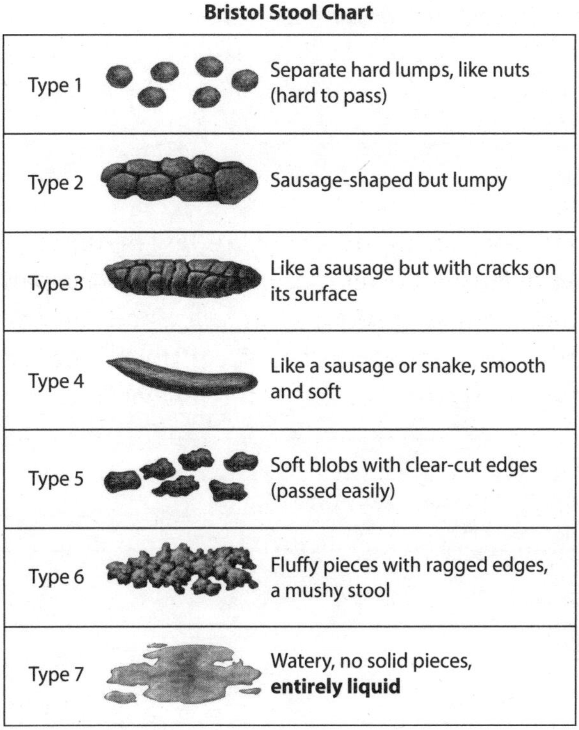

Bristol Stool Chart

Type 1		Separate hard lumps, like nuts (hard to pass)
Type 2		Sausage-shaped but lumpy
Type 3		Like a sausage but with cracks on its surface
Type 4		Like a sausage or snake, smooth and soft
Type 5		Soft blobs with clear-cut edges (passed easily)
Type 6		Fluffy pieces with ragged edges, a mushy stool
Type 7		Watery, no solid pieces, **entirely liquid**

Figure 7 The Bristol stool chart

Loud bowel sounds/borborygmi

These are the rumbling sounds we sometimes hear coming from our gut. It is normal to have these sounds from time to time, but in some people they can be very loud and very embarrassing. The technical name for these sounds is 'borborygmi'. These sounds are a sign that the bowel is contracting quite intensely and they can be accompanied by cramps.

Wind (flatulence or flatus)

This is another embarrassing topic for people to discuss. Passing wind from the back passage is entirely normal, and the average person passes wind ten to 15 times per day. Some people, however, feel that they are passing much larger amounts of wind than could be considered normal, or that it has an abnormally strong odour. This can be socially very embarrassing, particularly if controlling the wind is a problem.

Investigation of lower GI symptoms

Not every test is needed in everyone. This list is not exhaustive but would include the following:

Blood tests	Stool tests
Full blood count	Stool calprotectin (inflammation marker)
Full biochemical profile	Stool culture for bacterial infection or bacterial toxins
Coeliac bloods	Stool analysis for parasites
Thyroid function tests	Stool elastase (measure pancreatic enzyme levels)
Inflammation markers (CRP/ESR)	
Vitamin levels (B12, folate, vitamin D), ferritin and iron levels	

Colonoscopy

Before undergoing a colonoscopy you will have to drink a large volume of a liquid laxative to clear out the bowel. This preparation or 'prep' as it is called is probably the worst part of the entire procedure. A colonoscopy involves passage of a narrow tube with a camera at the end, into the anus and up into the top of the colon (the caecum). In many instances, the scope will be inserted into the lower end of the small bowel. Biopsies can also be taken. The procedure can be a little uncomfortable, as the endoscopist has to blow a little air into the bowel during the procedure. As a result it is generally done under mild sedation, but can be done without any sedation if you so choose.

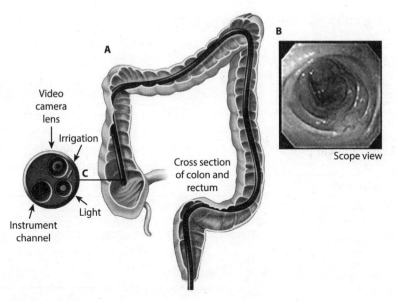

Figure 8 A colonoscopy

Scans/imaging

Depending on the symptoms, your doctor may think it necessary to arrange a CT scan of the abdomen, an MRI of the small bowel or an abdominal or pelvic ultrasound. A pelvic ultrasound is a good test to check for some common gynaecological problems such as ovarian cysts or fibroids. These can sometimes mimic gut symptoms.

What tests are necessary?

This will very much depend on your symptoms, your age and your doctor's evaluation. If you are a 30-year-old woman who has bloating and a several-year history of altered bowel pattern, then IBS is very likely. In this situation some blood tests and perhaps a stool calpro-tectin test to rule out inflammation (IBD) in the bowel would be appropriate first-line tests. If these tests are normal, then endoscopic investigation would not be necessary. A breath test for SIBO might also be considered.

On the other hand, if you are a 55-year-old woman and have developed new onset constipation and bloating, then you definitely will require a colonoscopy to rule out anything serious. You should also have the appropriate blood tests done.

Anna's story

Anna, an 18-year-old girl in her final year of school, was very conscientious, and put herself under a lot of pressure to do well in her exams. When Anna was referred to see Barbara, she had been having diarrhoea for three years. At the start this was two or three times per day, but over the last six months this had become worse and she was having four or five episodes of diarrhoea per day. This was interfering with her concentration in class, and she required frequent toilet breaks during the day. She admitted to feeling under a lot of stress with her studies, made worse by her diarrhoeal symptoms. Her mother felt that Anna had IBS and had done a lot of research online and had tried numerous elimination diets over the years, none of which had worked. During a visit to their family doctor a year previously, Anna had some bloods done and these showed a very slightly raised CRP (inflammation marker), which was put down to a viral infection that she had at the time.

Within a couple of minutes a few things became clear: (1) She was getting up at night with diarrhoea several times per week; (2) She had been getting recurrent mouth ulcers for about two years; and (3) She had lost about 7kg in the past year.

As we said before, getting up in the middle of the night to pass a bowel motion is extremely uncommon in IBS and is generally a sign of 'organic disease'. Additionally, mouth ulcers can occur in IBD and coeliac disease – and examination showed that Anna had two small ulcers in her mouth and that her lower abdomen was a little tender.

Based on this short history, Barbara was sure that Anna did not have IBS and that she more than likely had IBD. We repeated some blood tests, which showed that the CRP was raised once again, though this time there was no infection to explain it. She was also mildly anaemic. A stool calprotectin test was markedly raised, consistent with inflammation in the bowel. Anna underwent a colonoscopy that confirmed the presence of Crohn's disease. She was commenced on specific treatment for Crohn's disease, to which she responded very well with almost complete resolution of her symptoms within the next six months.

Take-home messages

- Every investigation is not required in every person and a lot of information can be gleaned from blood and stool tests.
- If you are younger, a colonoscopy is often not necessary to investigate bowel symptoms, but is definitely warranted if you are over 50 years of age.
- Diarrhoea waking you from your sleep (nocturnal diarrhoea) is very uncommon in IBS and usually indicates an organic condition – you should always discuss this with your doctor.
- Words to describe symptoms can mean different things to different people; your doctor should help tease this out with you.
- The Bristol stool chart is a very handy and easy-to-use way to describe your stools to your doctor.

4

The hidden world of your gut microbiota

Before we go further, there are a few headline things that you should know about the bacteria and other microorganisms in your gut, collectively known as the gut microbiota (GM). There are ten times more bacterial cells living in the entire human gut than there are human cells in our body! That is a staggering number of bacteria. The vast majority of the bacteria (approximately 70 per cent or 50-plus trillion) live in the colon, or large intestine.[2] The term microbiota is generally used to describe all of these bacteria, although it also refers to other microorganisms such as viruses and fungi that live mainly in the colon. The term 'microbiome' is often used interchangeably with the word 'microbiota, but strictly speaking they are different, and the word microbiome refers to all the genetic material (their DNA and RNA) carried by the microorganisms, rather than the actual microorganisms themselves.

What is a healthy gut microbiota?

How many times have you heard or read a claim that this probiotic or that food will nourish all the healthy bacteria in your gut? As the current state of knowledge stands, there is no one GM 'signature' that indicates that someone is healthy. However, we know that having lots of different bacteria is good – *plenty of diversity*.

Here are a few headline facts about your GM:

1. There are a lot of microorganisms inside us. Over 50 trillion bacteria live in the colon and the average person has between 300 and 1,000 different bacterial types in the GM, although 30 to 50 bacteria usually dominate in any one person. We also carry small amounts of fungus (around two thirds of people carry the fungus candida in their bowel), some viruses and other microorganisms. Many of the microorganisms have not been identified and so to some extent the

GM is uncharted territory! In particular the microorganisms that live in the small bowel are not well characterized and have been called the *'mysteriome'*.[2]

2. There is no one specific bacterial mix or pattern of GM, no 'GM signature' that indicates that someone is healthy or that the GM mix is a healthy one. Each individual has a different and unique GM mix. However, we know that having lots of different bacteria is good – this is called *microbial diversity*, and in general, the more of this, the better. You can achieve GM diversity by having plenty of variety within your diet. Reduced diversity and altered bacteria of the GM is known as *dysbiosis*.

3. Dysbiosis has been associated with many different medical conditions such as Crohn's disease, ulcerative colitis, IBS and even diabetes.[7] It is still not entirely clear if dysbiosis is the cause or the result of the particular medical condition and this is a topic of much current research.

4. Location, location, location! When we talk about gut bacteria, we want the right bacteria in the right place (colon). Problems arise when we have the wrong bacteria (dysbiosis) in the right place (colon) or bacteria in the wrong place (small intestinal bacterial overgrowth).

5. The GM are not passive bystanders in your gut; they are very active and produces lots of substances – chemicals and hormones – that interact with our bodies in hugely complex and poorly understood ways. The GM also produce many vital vitamins including vitamin K and the B vitamin group, as well as chemicals such as tryptophan (a precursor of serotonin, which is vital to our mental health), dopamine and short-chain fatty acids (SCFAs).

6. The GM of men and women are different.[8] The GM are involved in producing oestrogen, and we have oestrogen receptors in our gut. IBS is up to 2.5 times more common in women than in men, and it is fascinating to speculate that our differing GM make-ups might be contributing to this.

7. Menopause seems to affect the GM too.[8] There are differences in the GM of women before and after the menopause. Many women notice a change in their bowel functioning around and after menopause and it is interesting to think that this may be related to changes in the GM, among other factors.

8. Your diet has a *big* impact on the make-up of your GM. In general, diversity in your diet leads to diversity of your GM. Foods that alter the GM in a beneficial way are known as prebiotics. Fibre is one of the most important foods for your GM and has been shown to increase the number and the diversity of the GM.[9]

Where do the GM come from?

The foetal gut is relatively sterile and colonization of a newborn infant's gut with bacteria begins immediately after birth. In babies born by vaginal delivery, the colonizing bacteria come from maternal colonic and vaginal bacteria. In babies born by caesarean section the initial colonizers are from the skin and the hospital environment. These differences have led to theories about people who were born by caesarean section possibly being more likely to develop IBS. There is a lot of discussion on social media about this, and there is some interesting research but this has not been proven.

Breast-fed and formula-fed babies have different GM. Breast-fed babies tend to have a more limited range of bacteria. Regardless of the initial feeding, there is rapid change in infants' GM following weaning and introduction of solid food. By the time a toddler has been completely weaned from formula or breast milk, more adult-like GM patterns have emerged. Some people have suggested that breast-feeding is possibly protective against developing IBS in later life. As before, this is not proven.

What do the GM do?

Entire books have been written about the GM, so we can just give you a flavour of the known effects of the GM within our bodies. There are five main families (technically called *phyla*) of bacteria that make up the GM. About two thirds of the bacteria belong to a family called Firmacutes, and these contain lactobacilli, which have become a household name as they are used in many probiotics. The four other families contain other bacteria such as *bifidobacteria*, *clostridia*, *enterobacteria*, *Escherischia coli*, bacteroides to name but a few, as well as thousands of bacteria that do not yet have a name. Each of these different bacteria has a distinct metabolism and interacts with our gut in different ways.

Beneficial GM produce a number of substances called short-chain fatty acids (SCFAs) such as butyrate, propionate and acetate.[10] Butyrate nourishes the cells lining the colon. Propionate is absorbed from the gut and travels to the liver, where it plays an important role in glucose metabolism and might be involved in the prevention of Type 2 diabetes. Acetate is used as a food source by other bacteria and all of these SCFAs are thought to play a role in controlling our appetite. SCFAs also act as neurotransmitters in pain-pathways in the gut. Other GM convert tryptophan, an amino acid present in foods to serotonin (95 per cent of the serotonin in our body is produced by the GM), an important neurotransmitter that plays a vital role in the gut–brain axis and mental health, of which, we spoke earlier. The GM also produce important vitamins such as vitamin K and the B family of vitamins.

When we eat a diet high in animal fats, some GM produce substances that increase the risk of heart disease and stroke. There is some fascinating research suggesting that the certain types of GM cause us to crave certain foods. Imagine – it is not you who wants to eat that sugary cake or sweet, it is the streptococcus present in your mouth or your gut that is making you crave it...![12]

Every day we are learning more about the functions of the GM in our body and it is becoming abundantly clear, that a healthy and diverse GM is vital to our overall gut and body health.

Your individual GM

In any given person, the GM appears to be quite stable over time, but there is large variation among individuals and between people in different places. People living together in the same household tend to have similar patterns – this may reflect their similar diet, or the fact that bacteria may be passed from one person to another through close contact. The American Gut Project is a fascinating study that started about ten years ago.[11] People from all over the USA were given the opportunity to voluntarily submit stool samples for analysis of their GM. They also provided lots of information about their diet and other lifestyle factors such as smoking habits, alcohol intake and exercise. Some very interesting insights have come out of this and other research:

1. Plant-based foods, because they are high in fibre and fermented foods have a positive effect on GM. In fact this study found that the more plant-based foods that you eat in a week, the greater the diversity of your GM. People who ate more than 30 different types of plant-based foods were at the top of the class in terms of the diversity of their GM.

2. People who had sedentary lives, and who ate larger amounts of processed foods or animals fats and fewer than ten different types of plant-based foods per week had less diversity.

3. Many forms of alcohol reduce GM diversity, although red wine in moderation has been shown to increase it.

4. Smoking has a negative effect on GM diversity.

5. Exercise has a positive effect on GM diversity.

6. A gluten-free diet seems to cause a reduction in *bifidobacteria*, one of the beneficial GM (for this and other reasons we do not recommend strict gluten-free diets unless someone has coeliac disease or a definite wheat allergy).[12,13]

7. A diet low in fermentable carbohydrates, such as a low FODMAP diet also seems to reduce gut *bifidobacteria*.[14,15]

Fibre

We think of fibre as a 'super-food' for the gut, and are devoting an entire chapter to discussing its many and wondrous effects. But for now, understand that fibre is a food that is not absorbed by the small intestine, and so arrives in the colon in all its undigested glory, providing a literal feast for the GM in the colon. One of the downsides of this is that more gas can be produced.

Is there any difference in the GM of women and men?

It turns out that there is! The GM of women and men are quite different and these differences may be due in part to different dietary habits (men tend to eat more protein and to eat more of everything).[8]

Hormonal factors might also be playing a role in these differences. The differences in GM between men and women may also be a factor in IBS being more prevalent in women and why men and women with IBS often have different patterns of symptoms.

There are also differences in the GM in women before and after the menopause. The GM also seems to play an important role in production of female sex hormones.

Dysbiosis

We mentioned that GM diversity is the hallmark of a healthy gut and that reduced diversity, or overgrowth of different strains of bacteria in the colon is called *dysbiosis*.[16,17]

It is well established that patients with certain digestive diseases such as IBD or coeliac disease have altered patterns of bacteria within their GM. Basically, you could choose any disease nowadays, and someone has probably studied the GM of patients with that disease and compared this to healthy people. A common theme is emerging: patients with a given disease show evidence of an altered GM or dysbiosis.[18] However, we just need to be a little cautious in interpreting these results: in most cases we do not yet know if the dysbiosis is the cause, or the result of the disease (chicken or egg?).

The GM and irritable bowel syndrome

We will be discussing IBS in detail in the next chapter, but here we will just summarize some of what is known about the GM and IBS as follows:[7]

1. The GM in some, but not all, people with IBS differ from those of healthy controls.
2. There is no 'GM signature' that is found in all patients with IBS.
3. Some people with IBS seem to have less GM diversity or dysbiosis. This raises the possibility that increasing the GM diversity might be helpful in IBS.
4. Gases produced by GM digestion of certain foods (hydrogen, carbon dioxide, methane and hydrogen sulphides) as well as other

metabolites may contribute to the development of IBS symptoms. There is a large interplay between diet and the GM and the two things need to be considered together.

What about the fungi in the gut?

We are introducing yet another new term – the *myco*biota – which refers to all the normal, healthy fungi (plural of fungus) in your gut (like the bacteria, these live mainly in your colon, but also in your mouth). The gut is home to over 250 different fungi and they account for 0.1 per cent of the entire microorganisms in the gut. The most common of these is the *Saccharomyces* family (the yeast fungus used to make bread), which is present in almost 90 per cent of people, and next to that is candida, which is found in 60 per cent of people.[19]

IBS and candida

If you do an internet search using your 'IBS symptoms', possibly including fatigue and 'brain fog', you may be told that you have 'candida overgrowth'. Certainly there is evidence that candida levels are increased along with some other fungi in the faeces of some IBS patients.[19-21] But the changes are relatively small, and amount to only a part of the entire complex network of crosstalk between all the microorganisms in the gut. Just as with the bacteria story, it appears that it is good to have lots of different types of fungi in your gut – diversity is good.

It is overly simplistic to say that IBS is caused by candida overgrowth as the situation is much more complex than that. But it is likely that candida, as a normal gut inhabitant, may be playing a part in the big complex tapestry that is IBS. However, targeting treatment at candida alone is unlikely to be the answer.

What causes altered candida levels in the gut?

1. A course of antibiotics can lead to increased levels of candida in the mouth and gut. These changes are usually subtle but sometimes are enough to cause symptoms (candida in the mouth for example). This can be treated if causing symptoms, but even if left untreated, the candida levels will tend to re-set back to their 'normal' levels for that person.

2. More frequent brushing of the teeth has been found to reduce candida in the faeces, as a lot of the candida in the gut are thought to originate from the mouth.

3. Diet: high carbohydrate and high sugar diets have been associated with increased levels of candida in the faeces. Therefore the 'candida diet' advocates a low sugar and low carbohydrate approach. This diet may be beneficial to IBS symptoms in general – as we will learn in later chapters, too much of the wrong carbohydrates can be problematic to patients with IBS – but this is unlikely to be solely because of its effects on candida.[22]

What about candida antibody tests?

Occasionally patients come for a consultation and bring the results of these blood tests with them. The commonly used tests measure IgG antibodies to candida. Some of the tests measure three types of antibody (IgG, IgM and IgA) but these simply give evidence of exposure to candida (either right now or in the past). Anyone who has been exposed to candida (this will be most of us at some time in our lives) will develop antibodies to candida. We do not recommend the use of these tests.

Is there any value to a 'candida' diet?

The concept of 'candida overgrowth' in IBS has been taken out of context by popular media. Candida, along with the other fungi (saccharomyces being the most common) and the trillions of bacteria in the gut are all part of a complex ecosystem living inside of, and interacting with us. Singling out one organism, like candida, as being the main cause of IBS is misleading and quite simply, untrue.

Ultimately, a healthy GM, with both bacteria and fungi like candida, does not involve removing one type of bacteria or reducing one fungus such as candida. As we hope we have explained, it is all connected and a healthy, balanced diet achieves diversity of the gut microorganisms, not allowing any one organism to dominate.

How can you improve your GM?

There are a number of things that you can do to increase your GM diversity.

Top tips on how to improve your GM diversity

- Eat more plant-based foods, fibre and 'prebiotic' foods.
- Include fermented foods, which contain high levels of bacteria.
- Reduce your intake of highly processed foods, and saturated (animal) fats- we're big supporters of a Mediterranean-style diet, which ticks all the right boxes.
- Reduce your alcohol intake (a little red wine can be beneficial to your GM).
- Get adequate exercise and sleep.
- Do not smoke.
- Do not graze continually – the gut likes to get a rest from food – remember the MMC and the housekeeping wave.

IBS alert

While plant-based foods are rich in fibre and prebiotic substances, there can be some downsides to a high-fibre diet: increased gas and bloating (gases are produced by the GM during their fibre feast) can be a particular problem for those who have IBS. We recommend that each person should try to work out how much fibre she can comfortably tolerate and to increase fibre in the diet gradually. We will focus on this is more detail in Chapters 11 and 22.

Take-home messages

- The more the merrier and diversity is good.
- There is no GM 'signature' for IBS and, at this point, stool tests to analyse your GM have no clinical value.
- Your GM are extremely important to your overall health and wellbeing and have far-reaching effects beyond your gut.
- You can improve your GM diversity and number by eating a healthy diet, which contains plenty of plant-based foods and fibre.
- Your GM can also be nurtured by improving your lifestyle – cutting back on alcohol, and getting adequate sleep and exercise.
- Candida is not the cause of IBS and candida antibody tests are of no value.

Section 2

WOMEN'S DIGESTIVE PROBLEMS

5

Irritable bowel syndrome

What is irritable bowel syndrome (IBS)?

If you are reading this book, chances are that you have been diagnosed with IBS, and are frustrated by a difficulty in controlling your symptoms and the impact that this is having on your quality of life. About ten per cent of the world's population has symptoms of IBS and 65–75 per cent of them – i.e. approximately three out of four – are women.[23,24] This means that there are another 380 million women who may be suffering just like you are. You are not alone! The levels are higher in the developed world and younger people are more likely to have IBS than older people.

IBS is not new. What is new is that, thankfully, there is now a greater acceptance of IBS as a very real and incapacitating condition, and better acknowledgement by the medical profession and by society of the personal and societal impact of IBS. You have known first-hand for years the negative impact that this has had on your daily life, and yet you may have been told over and over again that there is 'nothing wrong' and that 'all the tests are normal'.

> 'IBS is very frustrating: it dominates life style and daily activities mostly through its unpredictability. You must always plan for the "what if" – "what if I eat more", "what if toilet facilities are not available". It inhibits your social life and sexual activity. IBS is frustrating and that's the bottom line.'

> 'The biggest problem is that no one (in the medical field) treats the whole person. I feel more like I'm going to a drug dealer than someone who looks at the problem in its totality. As a result I have turned my attention to helping myself, and have had some degree of success. I wish doctors would listen to patients more when we talk about the symptoms and how they affect our daily lives'.

These quotes are from two women with IBS who took part in a study a few years ago. Any woman who has been diagnosed with IBS will identify with these women. Many suffer in silence for years. In a large study published in 2009, 14 per cent of IBS patients said that they would take the chance of a 1-in-1,000 risk of death to receive a treatment that would free them from their IBS symptoms! In that same study, a large number of patients said they would give up 15 years of their life if they could be free of their IBS symptoms.[25] This gives us some idea of how severe such symptoms can be.

IBS is very frustrating and has a detrimental effect on women's quality of life. Many women with IBS have had unsatisfactory experiences with healthcare providers (mainly doctors) and find that those around them (family and friends) and the medical profession underestimate their symptoms. Not surprisingly many of those with IBS have problems with low mood and depression, which results in a vicious cycle with a downward spiral of mood and worsening digestive symptoms. Indeed, studies have shown that there are three recurrent themes that dominate the lives of those experiencing severe IBS;[26]

1. sense of frustration;
2. sense of isolation;
3. dissatisfaction with the medical/health system.

We explained earlier how IBS is a functional disorder of the lower GI tract. This means, disappointingly for many people, that there is no diagnostic test that will tell you or your doctor that 'yes, you definitely have IBS'. To be diagnosed with IBS, you need to have certain symptoms, for at least three months. Before a diagnosis of IBS is made, as we hope you now understand from the previous chapters, it is very important that you have had at least some basic tests to rule out the main 'organic' conditions that can mimic IBS. At the very minimum you should have had the basic blood tests done, and possibly some stool samples checked. Your GP/family doctor may also have referred you to a specialist and you may have had endoscopies performed. All such tests will show entirely normal results if you have IBS.

The Rome Foundation and the Rome IV criteria

Because IBS does not have a single diagnostic test, in the early 1990s a group of international experts established the Rome Foundation to try to draw up guidelines to enable physicians to diagnose IBS and other disorders. As a result of the accumulating evidence for the role of the gut–brain axis in these conditions, they are called 'disorders of gut–brain interaction' or DGBIs.[27] The Rome Foundation guidelines have been incredibly important in improving our understanding of DGBIs, in particular IBS and functional dyspepsia (FD). It has enabled further research to be done in the whole area, including much of the fascinating research on the microbiota, probiotics and diet, which have generated such huge interest in the media.

Rome IV definition of IBS

Criteria for diagnosing IBS

Recurrent abdominal pain for at least 1 day per week for the past 3 months
AND 2 or more of the following:

- The pain is related to defaecation (emptying the bowels)
- Associated with a change in frequency of stool (there may be constipation, diarrhoea or a mix of both)
- Associated with a change in the appearance of the stool

Strictly speaking, to be diagnosed with IBS, the symptoms must have started at least six months previously. However, in practice, this is not a hard and fast rule, and if you are a patient experiencing these symptoms, six months is a very long time to wait before someone is prepared to make a diagnosis. What's more, if you are having gut problems you certainly don't have to wait six months to seek help.

In IBS, abdominal distension or bloating is a very common symptom, but is not considered absolutely essential in order to make a diagnosis of IBS. In our experience abdominal bloating is one of the most problematic symptoms that many women with IBS complain of, and is often the main reason for seeking professional help. We will be speaking more about bloating in Chapter 21.

IBS subtypes

Based on these symptoms, a number of subtypes of IBS are recognized, and if you have IBS, you may recognize your own from the following descriptions. Classifying IBS into different subtypes is helpful, as treatment strategies will differ depending on the symptoms. If you experience severe bloating with diarrhoea, you may be helped by different measures from someone who has bloating with constipation and who opens their bowel only once per week.

IBS subtypes	
Subtype	Symptoms
Constipation-predominant IBS (IBS-C)	More than a quarter of the bowel movements are reported as being Bristol stool chart (BSC) types 1 or 2 (*see Chapter 3 for the Bristol stool chart*).
Diarrhoea-predominant IBS (IBS-D)	More than a quarter of the bowel movements are described by the patient as being BSC types 6 or 7.
Mixed or alternating type IBS (IBS-M)	Hard stools (BSC types 1 and 2) and loose stools (BSC 6 and 7) are equally common.
IBS unspecified (IBS-U)	The bowel habit cannot be accurately described by any of the above patterns.

Approximately 50 per cent of people with IBS fall into the mixed (IBS-M) or unspecified (IBS-U) categories at any one time. We also know that the condition can change over time, so that someone who is classified as IBS-M now may, in two years' time, have a more constipated bowel habit, and so fall into the IBS-C category. So, while classifying IBS in this way is helpful as a starting point, things can change.

IBS-D and bile salt diarrhoea/malabsorption alert

In Chapter 2 we briefly mentioned a condition called 'Bile Salt Diarrhoea'. This can mimic exactly the symptoms of IBS-D, often with urgent, loose stools, particularly triggered by eating. It is more common if you have had either your gallbladder removed in the past, or surgery on your small intestine, but it is also extremely common in people who have been diagnosed with IBS-D. In fact, almost one third of people diagnosed with IBS-D have been shown to have bile salt malabsorption. A really simple test for this is for your doctor to prescribe a 'bile mop' medication (your GP/family doctor can prescribe these medications). If your symptoms are being caused, even in part, by bile salt malabsorption, you will see a rapid improvement on these medications.

Diagnosis of IBS

While many people self-diagnose or see a complementary/alternative medicine practitioner who makes a diagnosis of IBS, it is vital that an appropriate assessment is done at the outset, or at some stage, to ensure there is no organic condition present that may be mimicking the symptoms of IBS.

There are essentially three or four steps involved in the diagnosis of IBS:

- a detailed history with a medical practitioner;
- a physical examination;
- appropriate laboratory investigations (as outlined previously);
- further investigations (in some cases only).

Amanda's story

Amanda, a 42-year-old mother of three young children who worked full-time as a nurse, had been suffering with chronic constipation for years, and in the past year had been passing a bowel motion only once per week. She also described lower abdominal pain, cramps and abdominal bloating. She had no 'red-flag' symptoms. She had never lost the 'baby weight' after her third child, who was now five years' old, and had joined a weight loss group about six months previously. Since then, she had managed to lose 8kg in weight through conscious effort, but her bowel function remained constipated and the abdominal pain and bloating had become worse.

Examination was normal apart from loss of muscle tone in the abdomen; very common after three pregnancies. Relevant blood tests were normal.

Amanda met the criteria for a diagnosis of constipation-predominant IBS (IBS-C). Strictly speaking there was no real reason to perform a colonoscopy, however Amanda herself was very keen to undergo colonoscopy as a cousin had been diagnosed with bowel cancer at age 54. We agreed therefore that a colonoscopy would be of value, to ease her concerns, and it was normal.

We explained to Amanda that the diagnosis was IBS-C and explained the nature of this condition and how tests are normal. We were greatly reassured by the normal results. We also explained to her that it was unlikely that she would pass a bowel motion every day but that this was not necessary or normal for everyone. With the right approach, it should be possible to increase her frequency of bowel motions and to significantly reduce the bloating and abdominal discomfort, which really were her biggest concern.

Amanda was then seen by Elaine and a detailed dietary history revealed that Amanda's fibre intake was poor (less than 10g fibre per day) but her fructose (fruit sugar) intake was high. Juggling the demands of work and family life meant that her diet was not always as healthy as it could be. Since she had joined the weight loss group, she had been eating a lot more fruit and, to a lesser extent, vegetables (you can throw some fruit into your bag with less preparation!). This coincided with the increased bloating and pain. Elaine instigated our FLAT Gut Diet. In particular she aimed to gradually increase Amanda's fibre intake up to 20g per day to start with, and to restrict some other potentially bloating foods. We also recommended that she try to incorporate a regular exercise programme and to make some time for herself (easier said than done with three young children!) and to try some daily mindfulness, even five minutes per day. In particular she should try some Pilates to increase her core muscle strength. The abdominal distension that was causing her so much distress was worsened by the reduced muscle tone in her abdomen after three pregnancies.

Amanda made good progress, and her symptoms improved gradually. Four months later she had lost a further 4kg in weight, and had significantly less bloating. She was working on her abdominal muscles and, although she only managed to get to a Pilates class once per week, she found it very good for her mental health, as well as for her core strength.

Take-home message While there is no magic-wand cure for IBS, management of symptoms requires dietary and lifestyle measures. Sometimes symptoms can be worsened by seemingly 'healthy' foods such as too much fruit that is very high in fructose.

What causes IBS?

This really is the million-dollar question! IBS is what is called a multifactorial condition. This means that in any one person it is likely that a number of factors act together to cause them to develop the symptoms of IBS.[4,26,28] No one factor alone is sufficient to cause IBS, and if you find someone trying to sell you a one-size-fits-all solution, do not believe it. The following factors have been identified as potential causes/contributors to the development of IBS.

Early life events

Stressful events in childhood may affect one's response to illness later in life. How illness and pain are expressed within families and cultures seems to affect the likelihood of developing IBS later in life. Salmonella infection in early life is a risk factor for developing IBS in adulthood. Birth by caesarean section, or lack of breastfeeding in infancy, have been mentioned as possible risk factors for developing IBS in later life but these have not been proven.

Psychosocial factors/stress and mood disorders

Factors such as stress have an effect on the gut–brain axis and affect pain experience and thus the severity of symptoms that patients with IBS experience. Individuals with IBS report more stressful life events than people who do not have IBS. What's more, mood disorders are much more common in people with IBS compared to healthy individuals.[29,30]

In some patients psychological factors seem to play a particularly large role in their symptoms, and some emerging research suggests that someone's psychological status might also determine their response to different therapies.[31,32] People who reported more anxiety and depression were more likely to require more treatments. Including psychological factors and assessment into the diagnosis of IBS and other DGBIs might help determine early on who might benefit from different treatments, such as psychotherapies, diet or different medications.

Previous infection

Prior infection, such as a bad bout of travellers' diarrhoea or a severe bout of gastroenteritis, is sometimes pinpointed as the trigger or beginning of IBS in some people. We have seen many people over the years who can say with certainty 'I was perfectly healthy until I had a bad bout of gastroenteritis while on holiday four-and-a-half years ago and I haven't been well since' or 'On the 3rd of March 2017 I got food poisoning from a take-away meal and I have had problems ever since'. If someone can pinpoint their symptoms to a very specific time and date, it is highly likely that they had an initial infective episode and that this then triggered the development of post-infection IBS. In fact approximately ten per cent of people diagnosed with IBS fall into this 'post-infection' group. Bacterial and viral gut infections tend to be short-lived and do not cause chronic symptoms. However certain parasitic infections can cause chronic infection and thus chronic symptoms; the most common ones are giardia or cryptosporidiosis. In this situation where symptoms start very abruptly, a stool sample should be sent to check for parasites. With so-called post-infection IBS, the symptoms can last for months or even years after the initial infection.

Lifestyle factors

Diet and exercise play a large role in digestive function. Exercise can increase gut motility and reduce stress through releasing endorphins. Endorphins are also natural painkillers and so can reduce the symptoms of IBS. Obviously, this does not work for everyone, but a regular exercise regime has been shown to help improve IBS symptoms.[33] We will be discussing the effect of food and diet on IBS in Section 4 of this book, and how the FLAT Gut Diet can help you.

Marie's story

Marie, a 45-year-old woman, was admitted to hospital with a very sudden onset of abdominal pain, cramps, high fever and severe diarrhoea. She was passing 15 to 20 bowel motions per day and there was blood in the stools. She had been out to dinner a couple of nights previously. On admission Marie had signs of an infection and her stool sample came back positive for *Campylobacter* infection. Marie was treated with rehydration and antibiotics and her symptoms gradually improved. She was discharged from hospital one week later.

When Barbara first met Marie, she had been re-admitted to hospital two weeks after the initial episode, as she was still feeling unwell. She had on-going diarrhoea, now passing five or six bowel motions per day, mainly after eating, associated with abdominal cramps and bloating. The blood in her stools had cleared. Her appetite was poor and she had lost 5kg since her discharge from hospital. She felt absolutely awful, had no energy and was very worried about herself. Blood and stool samples were repeated: they were all normal.

On examination, Marie had evidence of weight loss, and was clearly anxious, but otherwise everything was normal. Barbara explained to Marie that she had had an initial severe bout of gastroenteritis and now appeared to have developed post-infection IBS. Despite the normal investigations, given her age and weight loss, a colonoscopy was performed to rule out any organic cause; this was normal.

Marie was seen by Elaine and commenced the FLAT Gut Diet. This approach helped her symptoms significantly – particularly the bloating and diarrhoea. But over the next few months she continued to complain of abdominal pain and very low energy levels. She was unable to resume working, as she had a busy job that involved a lot of foreign travel and she simply was not fit for it. Barbara prescribed a low dose of a tricyclic anti-depressant (Amitriptyline – we will be discussing the use of these medications in Chapter 14 and Appendix 1) and gradually her symptoms and energy improved. She went back to work on a half-time basis. Nine months later she was feeling almost back to normal, but still had to be careful with her diet.

Take-home message A bacterial infection is generally acute in onset but relatively short-lived. Persistence of symptoms after a bacterial infection may be due to post-infection IBS. In many cases of post-infection IBS, the symptoms improve spontaneously over time, but this can take months to years, and in the meantime, can be helped by dietary and medical interventions.

Christine's story

Christine, a 54-year-old woman, developed acute digestive symptoms on the plane home from a trip to the Christmas markets in Europe. She felt acutely unwell and had vomiting, diarrhoea and abdominal pain. This lasted about 24 hours, and after that she did not pass a bowel motion for over two weeks. She had persistent abdominal ache/pain and felt unwell. She visited her family doctor who ran some blood tests and checked a stool sample for bacterial infection – all of these were normal. She was reassured by these tests, but over the next month, she continued to feel generally unwell. She returned to her family doctor a month later and told her that she still felt unwell and was constipated; she was passing a bowel motion only once every five to six days, whereas her normal bowel pattern was once every day.

She was referred to Barbara. Christine remembered eating some fishcakes that tasted a little off, at one of the Christmas markets. Her symptoms began about 14 hours later on the plane home and things had not returned to normal. Barbara felt that, as with Marie, this may be a case of post-infection IBS triggered by food poisoning, in this case from the fishcakes. The only slightly unusual thing was that her bowel pattern was constipated – in most instances, post-infection IBS leads to a diarrhoea-predominant IBS or a mixed type IBS.

Barbara repeated some blood tests, all of which were normal. In view of the change in bowel habit, Christine's age and the slightly atypical pattern of her symptoms, a colonoscopy was also performed. To Barbara's surprise this showed a number of unusual looking, small, slightly oval-shaped, slimy structures in her colon. These looked like parasites, in particular one that lives in fish called the Herring worm. Barbara suctioned up all the worms she could see up through the colonoscope (yes, this is a little gross) and sent them for analysis. They were confirmed as Herring worms, and after the colonoscopy, Christine was given a course of anti-parasitic treatment. Six weeks later her symptoms had much improved.

Take-home message There are always surprises in medicine! Although Christine's story sounded very much like post-infection IBS, the features were a little atypical. The Herring worm is found in uncooked fish and cases have been described in people who eat sushi and sashimi. If you love Japanese food (as we do), please do not be put off as it is very healthy and such instances are too occasional to worry about.

What are the mechanisms involved in the development of IBS?

The exact mechanisms involved in IBS are not entirely known, however, certain abnormalities of gut function have been identified that play a role in the symptoms of people with IBS.[30] These abnormalities are not routinely tested for (more for research at this point in time) and may not be present in every person with IBS.

1 Abnormal gut motility

There is a lot of evidence showing that IBS is associated with subtle abnormalities of gut motility. Women tend to have slower rates of gastric (stomach) emptying than men, and we also tend to have weaker peristaltic waves than men, which would explain why more women tend to have constipation-predominant (IBS-C) or mixed-type IBS (IBS-M) than men. In men, diarrhoea-predominant IBS (IBS-D) is more prevalent.

2 Visceral hypersensitivity

There is now very good evidence that patients with IBS have 'visceral hypersensitivity'. This means that their enteric nerves (the nerves in the gut, remember) are overly sensitive to normal stimuli. Levels of bowel distension cause pain in people with IBS that do not cause pain in those without IBS.

3 Abnormal immune responses in the gut

Our knowledge in this area is constantly evolving. While patients with IBS do not have easily detectable inflammation in the lining of their gut, like that found in patients with inflammatory bowel disease (IBD), research studies have shown subtle inflammation, not picked up on standard tests. This inflammation, often centred around nerve endings, may lead to increased sensitivity of the nerves in the gut wall.

4 Altered GM or dysbiosis

We discussed GM and dysbiosis (reduced number or diversity of the GM) in Chapter 4. Studies on the microbiota and IBS began 15–20 years ago and much (often conflicting) information has emerged since then. However, there is a consistent finding that many patients with IBS have dysbiosis, with different patterns of bacteria seen in patients who have IBS-D compared to IBS-C.[19,34] The exact significance of this is not known at this time – is this a chicken or egg situation? Is the dysbiosis the cause or the result of the IBS? Watch this space.

5 Altered response to certain foods/food intolerance

We will be discussing dietary management of IBS in a lot more detail later in Sections 4 and 5, and will just mention it briefly now. Anyone with IBS will know that there are certain 'trigger' foods that can bring on or worsen their symptoms. There is so much information (and misinformation) available on the internet nowadays that most people with IBS will have tried excluding many different foods in an attempt to control their symptoms. Often one food group is excluded, followed by another, and another, and yet the symptoms persist! A study some years ago showed that 12 per cent of IBS patients studied had an inadequate dietary intake – in other words, their diet was deficient in important vitamins and micronutrients – to maintain health.[35]

In recent years, much evidence has emerged supporting the low FODMAP diet (we discuss FODMAPs in detail later in the chapter on food intolerance) as a dietary treatment for IBS.[36,37] We have also used the low FODMAP diet in thousands of patients, but have felt, as have other experts, that while it is effective in the short term, it can be too restrictive, and following it properly can be very difficult.[38,39] For this reason, our own practice has evolved to the FLAT Gut Diet. This reduces all the main trigger foods, but in a less restrictive way, and allows you to reintroduce them in a controlled way, to find your individual level of tolerance. Section 5 is devoted entirely to the FLAT Gut Diet, and we hope that you're going to give it a try.

6 Alterations in the gut–brain axis

We now know that patients with IBS and functional dyspepsia have alterations in the gut–brain axis (see Figure 4).[40] IBS patients have increased activation in parts of the brain involved in pain perception, in response to relatively minor changes within the gut. The end result of this imbalance is that that pain coming from the gut in patients with IBS is then amplified at the level of the brain, which in turn leads to distress and anxiety, and this in turn worsens the pain perception in the gut.

These six, interconnected mechanisms play a role in triggering and perpetuating the symptoms of IBS, but also of other DGBIs such as functional dyspepsia.

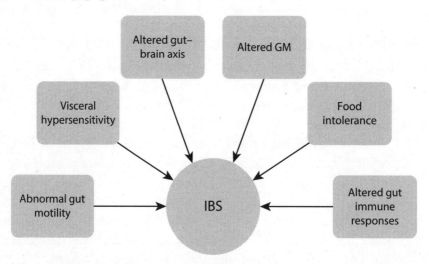

The end result of these changes is the vicious cycle of events, which we mentioned before:

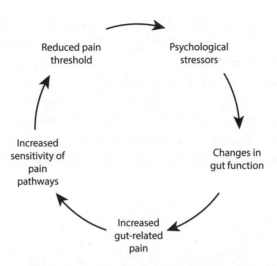

All of this leads to pain and discomfort, reduced quality of life, distress and increased anxiety. Anyone with IBS will identify with this.

We will show you how to manage these symptoms, when to seek help and how to empower yourself to regain control of your life.

How do you manage IBS?

As you will understand by now, no two people with IBS are exactly the same. If you have mild symptoms then some simple dietary and lifestyle changes may be enough to get them under control.

In Chapter 14 we'll show you the first three steps of the Gut Experts' 4-Step IBS Solution. If these measures do not help your symptoms enough, you will be shown Step 4 – the FLAT Gut Diet, a specific diet designed for IBS. All of Section 5 (Chapters 17–23) is devoted to the FLAT Gut Diet, which we are very excited to present to you for the first time in this book.

It is important that you understand at the outset, that while dietary management is a vital part of your IBS management plan, for many people it is not enough by itself. If you have had years of pain and bloating, coupled with the stress and anxiety that goes along with this, the 'nerve endings' in your gut are likely to be hypersensitive. What's more, bloating is a complex issue, which we discuss in detail in Chapter 15, and is not just caused by what you eat. For these reasons, you need to address all of these factors that are contributing to the vicious cycle of pain and discomfort.

If you have severe IBS, to truly get the best results, you need an individualized management plan, where all your triggers are addressed. It's important that mental health is nurtured and that significant stress and anxiety are treated, through whatever means is most effective. Visceral hypersensitivity may require treatment with a prescription medication to break the cycle of pain. We discuss all of this is more detail in Chapters 14 and 23, where we look at more specific management of IBS symptoms and also our TEAMS approach.

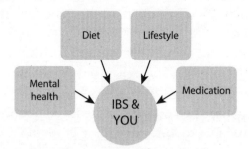

It is important to discuss both over-the-counter and prescription medication with your doctor, in tandem with all the other things you are working on with diet and lifestyle. We include some of the many medications used to treat IBS in Appendix 1.

Take-home messages

- IBS is much more common in women than in men and almost one-in-six women have it.
- IBS is much better understood than before, but there is still a lot more to find out.
- There are different types of IBS, diagnosed on the basis of the pattern of your bowel movements.
- If you have IBS, standard tests will be normal.
- One size does not fit all when it comes to treatment, and your treatment should be personalized to you.
- IBS is called a 'disorder of gut–brain interaction' but do not let anyone tell you that it is imagined or 'all in your head' – it is very real and can be extremely debilitating.
- There have been major advances in our understanding and treatment of IBS and you should not suffer in silence.
- If you have IBS-D, ask your doctor to consider giving you a trial of a 'bile mop' medication – the results can be quick and life-transforming.

6

Functional dyspepsia

Almost everyone has heard of IBS. When a patient is told that they have IBS, they understand what this means – they have heard the term, some of their friends may have it, generally they will have looked it up online, and the condition is now well recognized in society – as a result, the diagnosis is easier to accept as it is familiar. However, this is not the case for functional dyspepsia.

A relatively unknown and under-recognized condition

Functional dyspepsia (FD) is not known in social media or the mainstream press, and is poorly recognized outside the field of gastroenterology.[41] In fact, it is also unfamiliar to many GPs/family practitioners and we find that people are often a little reluctant to accept the diagnosis, as it is so unfamiliar. However it is time for FD to take its place on the main stage; people need to be aware of this common condition, which causes symptoms in hundreds of millions of people worldwide. When we are explaining FD to patients for the first time, we often describe it as 'irritable stomach', or that it is the upper digestive variant, or sister condition, of IBS. Because this is essentially what it is! It would have been easier to call this 'irritable stomach' but as things stand, we are stuck with the term functional dyspepsia (FD).

Who gets functional dyspepsia?

FD affects approximately one in five adults, making it even more common than IBS.[42,4] FD is also more common in people who have IBS, and it's thought that up to half of people with IBS also have FD.[43] This makes sense when you think of it, as the digestive system is a continuum from the mouth to the anal canal, and the same gut–brain axis controls the entire area. So, if there is an imbalance affecting one part of the digestive system, it might also affect other areas. Unlike IBS, FD appears to be more-or-less equally common in men and women, being only slightly more common in women. *It is estimated that over 800 million people worldwide have FD.*[24] In our own practices in recent years, we are diagnosing more FD in young women and girls in their teens and, while the cause for this is not clear, we think that it is likely related to increased stressors and also possibly stress and peer-pressure regarding food.

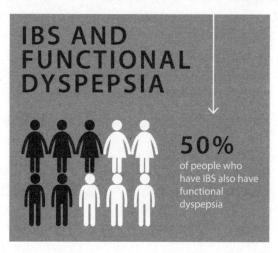

IBS AND FUNCTIONAL DYSPEPSIA

50%
of people who have IBS also have functional dyspepsia

The Rome Foundation and the Rome IV criteria

The Rome Foundation Group has helped classify FD more accurately, just as they have done for IBS over the past few decades. The big challenge ahead in the coming years is to get FD recognized and accepted in society, in the same way as IBS.[41]

What are the symptoms of functional dyspepsia?

Any one of the following:
- A feeling of unpleasant/bothersome fullness after eating – this is called post-prandial fullness (*enough to impact on normal activities*).
- Feeling full after eating a small amount of food. This is called early satiety (*enough to prevent you finishing a normal-sized meal*).
- Pain or burning in the upper abdomen/epigastric area (*enough to impact on normal activities*).*

AND
- No structural or organic disease that is likely to explain the symptoms.

AND
- The symptoms must be present at least three days per week for the past three months, and to have started at least six months ago.

Other common symptoms include:
- Nausea, excessive belching or feeling bloated after eating.

*The epigastric area is the upper, central part of your abdomen, just below the breastbone.

Functional dyspepsia subtypes

Just as with IBS, FD has been sub-classified into different types, based on the pattern of symptoms. There can be an overlap between the two subtypes, and many people will have features of both. In the first subtype, the symptoms are mainly related to meals, whereas in the second they are not, although in reality, many people with FD have an overlap between the two.

Functional dyspepsia subtypes	
Subtype	**Symptoms**
Post-prandial distress syndrome (PDS)	Bothersome post-prandial fullness (severe enough to impact on normal activities)
	Bothersome early satiety (severe enough to prevent finishing a normal-sized meal)
	These symptoms are mainly brought on by eating.
Epigastric pain syndrome (EPS)	Bothersome epigastric pain (severe enough to impact on normal activities)
	Bothersome epigastric burning (severe enough to impact on normal activities)
	In this subtype of FD, unlike PDS, the upper GI symptoms are not so strongly (or at all) related to meals.
In both types of FD other common symptoms include nausea, post-prandial bloating or excessive belching.	

We have all experienced these sorts of symptoms from time to time, maybe after a late meal accompanied by a few drinks, or a particularly heavy meal or after a bout of gastroenteritis, but thankfully these are usually short-lived. To be diagnosed with FD, these symptoms have to be present several days per week for a few months: the hard and fast rule is that they should be present at least three days per week, for at least three months, and should have started more than six months previously.

For some people the symptoms are a minor inconvenience – a bit of bloating after eating or some occasional nausea. However, for others the symptoms can be very problematic and have a hugely negative impact on their quality of life.

Most of the symptoms of FD occur after eating (post-prandial), although the pain in the upper abdomen may be more constant. If you get unpleasant symptoms whenever you eat, then it is not surprising that you might start to dread eating, and to worry about these symptoms coming on, even before you eat anything at all. This is called 'symptom anticipation'. This leads to a whole additional level of anxiety, which can lead to people avoiding food. When teenage girls and young women start to avoid food because they have these symptoms it can be challenging to be sure that this is not the beginnings of an eating disorder.

Ruling out other conditions

The symptoms we mentioned – bloating, a burning feeling, nausea, feeling very full – can also be caused by other 'organic' digestive problems such as a hiatus hernia or acid reflux, an ulcer or even gallstones or coeliac disease. Because FD is not well recognized even by doctors, many people are misdiagnosed as having acid-related conditions and are being given PPI medications, only for these medications to be ineffective. Therefore it is important to have appropriate tests done before a diagnosis of FD can be made, and this should be discussed with your GP/family doctor in the first instance. They might request some blood tests, refer you to a specialist for an upper digestive endoscopy (a gastroscopy or OGD) or arrange an abdominal ultrasound. When you have FD, all of these tests will be normal.

How long does FD last?

Just like IBS, for some people the symptoms of FD can come and go, particularly at times of stress, but for other people, the symptoms can be of much longer duration, or even life-long.

What causes functional dyspepsia?

All functional conditions, or DGBIs, involve a complex interplay between the mind and the body, and there is generally no 'one cause'. FD shares many common causes/mechanisms with IBS[4,44,45] and we are still learning.

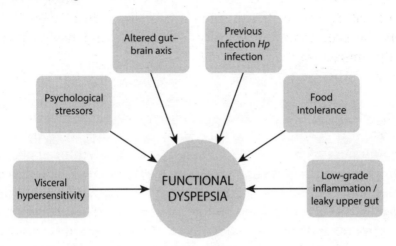

Figure 9 Mechanisms involved in the development of functional dyspepsia

Other causes of FD

A number of other changes have been described in people with FD:

- Poor relaxation (accommodation) of the stomach in response to a meal. The stomach is like a big sac, and it's supposed to stretch and relax when we eat. This doesn't happen properly in some people with FD, which is why they feel so bloated and full after eating.

- About one third of those with FD have slow emptying of the stomach, and a much smaller percentage have rapid stomach emptying. Both of these can lead to fullness, pain and discomfort – in one case the food is sitting in the stomach for too long and in the other, the food is entering into the duodenum (small intestine) too quickly. In the USA many people with FD may have previously been diagnosed with 'gastroparesis', which literally means paralysis of the stomach, or in other words, very slow gastric emptying. Nowadays many of these people would be re-classified as FD.

- Hypersensitivity of the stomach and duodenum to distension, acid and other chemical stimuli (these could be present in food)
- *Helicobacter pylori* infection.

Katie's story

Katie was a 17-year-old girl when Barbara first met her. Her family doctor had referred her to the clinic with symptoms of poor appetite, weight loss and nausea. She had lost about 7kg in weight in the previous six months and now weighed 50kg, with a BMI of 17. Her GP had done the appropriate blood tests and these were all normal. Both of Katie's parents accompanied her to the initial consultation, and they were clearly very concerned about the weight loss.

Taking a detailed history, Katie described a relatively rapid onset of symptoms of early satiety, post-prandial fullness, upper abdominal bloating and nausea about eight months previously. As a result of these symptoms, she had been eating a lot less than normal. There was no obvious infective trigger and the symptoms started during the summer holidays when she was not feeling under stress. Katie's mum said that Katie had always been slim, but that she had always been a good eater and had never seemed concerned about her weight. Katie herself felt that she was now too thin, although she was not overly concerned about this. The physical examination was completely normal.

Barbara felt that Katie's symptoms were very typical of functional dyspepsia, of the post-prandial distress syndrome (PDS) type. However, in a young woman of this age, one would also consider the possibility that she had developed an eating disorder and was using physical symptoms as an excuse to avoid eating.

The likely diagnosis of FD was explained to Katie and her parents and they were a little sceptical about it. It was clear that they were concerned about either an eating disorder or some serious organic condition. Barbara explained that at her age, an endoscopy would not generally be warranted based on these symptoms, and that instead we would perform some non-invasive tests to rule out other causes (breath test for *Hp* infection, blood tests for coeliac etc). Some dietary approaches to management were discussed – eating little and often, avoiding cold fluids and spicy foods – and Katie was put on a trial of domperidone to be taken before meals to see if this helped things. Katie and her parents were also given some information to read about FD.

Katie returned for review a few weeks later. The breath test and blood tests were normal but she was not feeling any better. It seemed important at this point to reassure Katie and her parents that there was nothing serious underlying her symptoms and an OGD was performed – this was completely normal, as were all the biopsies. Katie was started on a medication that can help with gastric relaxation and which has been shown to be helpful in FD. She was also started on dietary supplements to help her gain some weight, and she was referred to Elaine for a full dietary assessment.

Two months later Katie returned for review and was significantly better. She had gained 5kg in weight and her symptoms had improved immensely. She was still unable to eat a large meal, but this was improving.

Take-home message FD is a common condition. It can be difficult for people to accept, as it is not well known. In young people, particularly young women, there is often concern about a possible underlying eating disorder and this may need to be explored.

How do you manage FD?

Just as with IBS, the treatment of FD must be personalized. Treatment also needs to be holistic and address all aspects of the condition. Some aspects of FD respond less well to diet than others, particularly the pain and burning symptoms, and certain medications can be extremely helpful in this context. While it may seem preferable to manage things without medication, don't rule it out, as the right medication can make a big difference to how you feel. We discuss this in detail in Chapter 16 and Appendix 2.

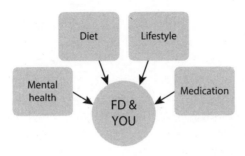

Take-home messages

- FD is very common and affects one in five adults at some point in their lives.
- We need to increase awareness about FD.
- FD and IBS share many common causes.
- Half of people with IBS also have symptoms of FD.
- One size does not fit all when it comes to treatment and your treatment should be personalized to you.
- Like IBS, FD is a 'disorder of gut–brain interaction' – it is very real and can be extremely debilitating, causing food-related anxiety and 'symptom anticipation'.
- We are learning more and more about FD, and the more we know, the better our treatment and solutions.
- Go straight to Chapter 16 now if you want to read about how to manage FD, without any further delay.

7

The effect of female hormones on gut function

Introduction

If you are a woman with IBS, you probably know that your GI symptoms can fluctuate dramatically over the monthly course of your menstrual cycle. This also seems to be the case for some women with FD.

The two main functional GI disorders, IBS and FD, are rarely seen in childhood, and tend to develop after the onset of puberty, particularly in the late teens and early twenties. They continue to be problematic for many women for the next 20 years or so (that's a very long time!) and then tend to become milder and less common after the menopause. This observation has led to interest in the role of the reproductive hormones, namely oestrogen and progesterone, in the development of symptoms in both IBS and functional dyspepsia.

The role of sex hormones on the female digestive system

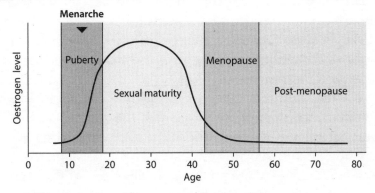

Figure 10 Oestrogen levels over the course of a woman's life

The diagram shows the average levels of oestrogen in a woman's body at different times in her life. You can see that there is a rapid rise in oestrogen levels during puberty, then sustained high levels during the reproductive years, before levels start to decline during the peri-menopausal years, until eventually levels become very low in the post-menopausal period of life.

Hormones, like oestrogen, progesterone, testosterone, thyroid hormone, cortisol (the stress hormone), and many, many more, all act by binding to specific receptors on and inside cells in the body. The binding of the specific hormone to its receptors sets off a cascade of activity and changes within the cell, which ultimately brings about the effects of that hormone. Not surprisingly, oestrogen receptors are found on the female reproductive organs, and bring about the changes involved in sexual maturation during puberty, the menstrual cycle and also the profound changes that occur in the body during pregnancy. However, oestrogen receptors are also found in every part of the body including the gut (particularly on the nerves of the gut), the brain and the spinal cord.[46]

IBS and the menstrual cycle

If you have IBS, you may already be aware that your IBS symptoms vary at different times of your monthly cycle. Many women find that their symptoms get worse just before and during their period. Some women will also find that their symptoms get worse around ovulation, only to improve again afterwards.[47] Women who do not have IBS also often find some changes in their bowel function over the course of their monthly cycle, but these changes are much milder and much less problematic.

The diagram in Figure 11 shows the slightly complicated monthly fluctuations in female hormone levels over the course of one monthly cycle. LH (luteinizing hormone) and FSH (follicle stimulating hormone) are hormones produced by the pituitary gland in the brain that control the release of oestrogen and progesterone. Day one of the cycle is the first day of the period. The first half of the menstrual cycle lasts about 13 to 15 days and is called the follicular phase. The upper line shows

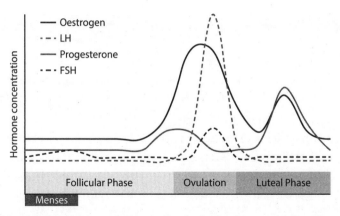

Figure 11 Monthly fluctuations in hormone levels during the menstrual cycle

oestrogen levels and the second line down shows progesterone levels – they are both quite low initially and then rise (particularly oestrogen) just before ovulation. Both then fall a little, only to rise again in the second half of the cycle, which is called the luteal phase. At the end of the luteal phase, usually around day 27 to 29 (of a 28- to 30-day cycle), the levels of oestrogen and progesterone fall rapidly.

If we look at the graph in Figure 11 again and now superimpose the times when IBS symptoms are at their worst, we see that just before ovulation, oestrogen levels are high, and at the end of the cycle, just before a period, oestrogen and progesterone levels are high and then fall very rapidly, just as a period starts. So, what are these hormones doing at the level of the gut?

How oestrogen and progesterone affect the digestive system

To say that the effects of these hormones on the gut are complicated is a bit of an understatement! The story is not a simple one, and a lot of the changes brought about by both oestrogen and progesterone are very complex, and the balance of these changes in any individual woman ultimately determines how her symptoms behave at different stages of her cycle.[47,48]

We have come up with our 'Top Ten' effects of what we know sex hormones can do to the gut, but just be aware that even this is just a taste of what is known.

The 'Top Ten' ways that sex hormones can affect your gut	
Effect on the gut	**Effect on IBS and FD**
Oestrogen slows down gut motility. This means that higher levels of oestrogen may reduce diarrhoea, or worsen constipation. The effect will differ in different people, and may depend on a person's predominant symptoms.	IBS-C may get worse, and IBS-D may improve when oestrogen levels are high.
Progesterone also slows down gut motility – this can result in increased constipation. It also slows down emptying of the stomach.	IBS-C may get worse and IBS-D may improve when levels are high. Worse FD symptoms in the second half of the cycle.
Oestrogen can increase mast cells (a type of inflammatory cell) in the gut wall, particularly around the nerves within the gut wall. This can lead to an increase in pain perception.	Worsening of IBS and FD symptoms when oestrogen levels are high.
Oestrogen increases serotonin levels in the brain. Serotonin is one of our 'happy hormones' and increased activity in the brain improves mood. Increased serotonin in the brain also inhibits pain pathways, and reduces central (brain level) perception of pain.	The balance of the effects of oestrogen on pain pathways can vary from person to person. In some, the effects of serotonin may outweigh the effects on mast cells, for example.
Oestrogen also increases serotonin activity in the gut. There are a number of different types of serotonin receptors on the cells of the gut; some act to increase pain perception and some reduce pain.	The balance of these effects will differ in different women, although on the whole, increased serotonin activity at the level of the gut will increase IBS symptoms.
Women with IBS have been found to have a higher concentration of oestrogen receptors on the cells of their gut, than those without.	This might partly explain why women with IBS are more sensitive to hormonal changes than their counterparts who do not have IBS.
In the second half of the cycle, the luteal phase, oestrogen seems to increase sensitivity to psychological stressors through alteration of cortisol receptors (natural stress hormone) in parts of the brain. We think that most women can identify with this!	Likely to worsen IBS symptoms.

The 'Top Ten' ways that sex hormones can affect your gut	
The GM of men and women are different AND, what's more, the GM in the male and female digestive systems are actively involved in the production of sex hormones including oestrogen, progesterone and testosterone.	IBS is 2.5 times more common in women. Interesting...?
Certain GM can reactivate oestrogen that is being cleared from the body into the gut (via the liver and bile). GM diversity and mix will affect how much oestrogen is re-activated and can thus influence oestrogen levels in the body.	Your GM mix and diversity is thought to play a role in many different medical conditions that affect women, including IBS, endometriosis, polycystic ovary syndrome (PCOS) and a number of different cancers that are related to oestrogen levels (breast and womb cancer).
Reduced sex hormone levels during and after menopause will reduce all the effects of these hormones on your gut.	Many women with IBS will find that their symptoms improve after menopause (although for some, other gut symptoms can become a problem).

What does all this science mean?

We have just described a lot of different and complex effects of oestrogen and progesterone on the female digestive system, and this is just a selection of some of the research that has been done in this area. At the end of the day, these simply explain the basis of what we already know – that IBS symptoms fluctuate during a woman's monthly cycle and that this is not imagined.

For most women we see as patients it is at the late phase of the cycle and at the beginning of menstruation that most problems arise – increased bloating, often increasing constipation, with diarrhoea then developing at the start of the period. There is increased pain, and add in some increased stress also, and it really is a recipe for significant discomfort, stress and feeling awful. The anticipation of this pattern repeating itself month-in-month-out is also very stressful, and women find themselves avoiding social situations, avoiding meetings or presentations in work, because they know that they will be feeling unwell.

Can anything be done to ease the monthly cycle-related IBS symptoms?

There is no 'one-size-fits-all' cure for the severe symptoms that some women with IBS experience over the course of their monthly cycle, but there is light at the end of the tunnel, and there are some simple measures you can take to help get control:

Six strategies to relieve period-related IBS symptoms

- **Knowledge is power!** Knowing your body and understanding why you are experiencing increased problems at certain times of the month is helpful. This is obviously not a cure, but it is important for your own self-esteem and confidence to understand that these are real changes and to anticipate them.
- **Check your fibre and fluid intake:** if you know that you always become more constipated before your period, and you are prone to constipation in general, then you need to pay particular attention to your dietary fibre, fluid intake and perhaps take some additional fibre supplements around this time.
- **Reduce your intake of bowel stimulants:** similarly, if diarrhoea is a more pronounced problem for you at a particular time of the month (in many women, this is often around period time), then you can try to anticipate this, and reduce fibre intake, reduce bowel stimulants such as caffeine and spicy foods and adhere more rigidly to the dietary regime that is right for you.
- **Modify your diet:** dietary modifications, such as the FLAT Gut Programme that we will be discussing in Section 5 can also significantly reduce bloating, pain and discomfort.
- **Try to control your stress levels:** other strategies might be particularly important at certain times of the month, including mindfulness and other means of relaxation. It is also vital to try to ensure that you get adequate sleep. It may involve some trial and error, but ultimately you should try to find the form of relaxation strategy that works best for you!

- **Consider artificial hormones or contraception:** the combined oral contraceptive pill (OCP) stops the fluctuations in oestrogen and progesterone over the course of the month. There is good evidence that the OCP can be very helpful, although it may not be the right treatment for everyone. Women who take the week off their OCP as prescribed and have a withdrawal bleed, may continue to get symptoms, although perhaps not as severe as when menstruating naturally. If the severe symptoms continue around menstruation, then there is the option of taking the OCP back-to-back (not taking a break of a week each month), and taking a break less frequently. Obviously, this should always be discussed with your GP/family doctor.

Endometriosis Awareness Alert

Endometriosis is a condition that affects up to one in ten women and is associated with abdominal pain, particularly around menstruation. It can cause painful or heavy periods and other symptoms, such as painful intercourse, bloating and cramps, some of which can mimic IBS. It is caused by womb lining-type cells (endometrial cells) developing outside of the womb, particularly around the Fallopian tubes, ovaries and in the pelvis, sometimes around the rectum. Because it can cause IBS-type symptoms, women are sometimes referred to a gastroenterologist for investigation. It can be difficult to diagnose as often it does not show up on scans.

If you experience severe period pain or other period-related symptoms, you should mention it to your GP/ family doctor.

8

Pregnancy and your gut

During pregnancy, the female body undergoes many physiological changes, all of which are designed to create a safe and nourishing environment for the growing foetus. Many of the changes that occur during pregnancy can affect gut function, both directly (there's a baby growing right in the abdomen exerting pressure on the digestive system) and indirectly, through hormonal effects. The main digestive symptoms that cause women problems during pregnancy are heartburn, acid reflux and constipation, and these are very common.

Over two-thirds of people with IBS are female. What's more, they are largely young females, which means that managing IBS can be a particular problem for many young women during pregnancy.

There are a number of other pregnancy-related conditions that also affect the gut, such as hyperemesis gravidarum (severe, intractable nausea and vomiting), but we're not going to discuss this here, as it's a specialist obstetric condition.

Direct anatomical effects

The foetus growing in the uterus can exert pressure on the rectum, which lies just behind the uterus, and is thought to contribute to constipation. Over a third of women experience constipation during the first and second trimesters and this actually tends to improve in the third trimester.[51] The fact that constipation often improves later in pregnancy suggests that hormones are playing a larger role than the mechanical effects of the enlarging uterus. Later in pregnancy, the uterus expands up into the abdomen and this pressure also contributes to acid reflux and heartburn. Over half of women experience acid reflux during pregnancy [52] and this usually starts at the end of the first trimester.

Hormonal effects

Many of the digestive symptoms that occur during pregnancy are now known to be caused by the myriad of hormonal changes that occur, rather than by the direct mechanical effects.[53] Think of the number of hormonal changes involved in the menstrual cycle and multiply that by ten!

Here are just some of the effects of pregnancy hormones on gut function:

- Progesterone slows gut motility, contributing to constipation, delayed gastric emptying and reflux.
- The hormone hCG (which is produced by the placenta and is the basis of pregnancy tests) makes the migrating motor complex (the MMC) less effective, and this can also contribute to constipation.
- Relaxin, another pregnancy hormone, reduces bowel contractions, and is yet another contributor to constipation and also acid reflux.
- The placenta produces a hormone called gastrin, which stimulates acid production by the stomach and may contribute to the development of acid reflux.
- Oestrogen affects pain perception and the functioning of the gut–brain axis, and changes during pregnancy might contribute to worsening IBS symptoms.

IBS and pregnancy – what happens?

Despite both IBS and pregnancy being so common (!), there is a lack of research that considers the effect of IBS on pregnancy and that of pregnancy on IBS. Many medical studies actively exclude pregnant women, for safety reasons, and so much of what we know about IBS and pregnancy is based purely on observation. And however little we know about the relationship of IBS and pregnancy, we know even less about functional dyspepsia and pregnancy.

The effect of pregnancy on IBS varies from woman to woman. Some women will say that they feel a lot better during pregnancy, whereas others find that the symptoms become more problematic than ever. This likely reflects whatever the underlying symptoms might be; if you have IBS-D (diarrhoea-predominant IBS), then all the hormonal effects, leading to relative constipation, may in fact greatly help your symptoms. If, on the other hand, you have IBS-C (constipation-predominant IBS), then that may worsen during pregnancy. If bloating has been a significant problem, then that too may worsen during pregnancy. Some women find pregnancy a stressful time, for many reasons – worrying about a successful outcome, juggling many demands – and this can alter the functioning of the gut–brain axis, leading to increased IBS symptoms.

Managing constipation during pregnancy

Over a third of women have constipation during the first and second trimesters; this reduces to a fifth in the third trimester.[51] Managing constipation during pregnancy has much in common with managing constipation in general, and we talk about this a lot more in Chapter 19. However there are a few important considerations:

Five tips for coping with constipation during pregnancy

- **Diet matters:** fibre and fluid intake are very important, so plenty of plant-based foods.
- **Gentle laxatives are fine:** fibre supplements such as psyllium are safe, and where needed other agents such as magnesium salts, lactulose or polyethylene glycol are also safe.
- **Stronger laxatives are not:** stimulant laxatives such as senna and bisacodyl should be avoided as there are concerns that these might trigger premature labour.
- **Enemas should be avoided:** phosphate enemas should not be used during pregnancy, as they may affect bone formation in the foetus.
- **Exercise is helpful:** it is important to keep as active as possible and to get regular exercise during pregnancy, for many reasons. This also helps constipation.

Managing IBS during pregnancy

Most of the general measures and treatments for IBS are similar for pregnant women, apart from the use of medications. Many of the newer specialist medications used to treat IBS-C or IBS-D, which we mention in Chapter 14 and Appendix 1, do not have sufficient evidence and data to support their safe use during pregnancy and should be avoided. If you become pregnant while on one of these medications, you should stop taking it until you can discuss further with your obstetrician or gastroenterologist.

Peppermint oil is very safe during pregnancy as an antispasmodic, although it can worsen acid reflux symptoms.

Some medications used to treat more severe IBS, such as tricyclic antidepressants, appear to have a low (but not no) risk of causing foetal abnormalities and should be discussed with your doctor if you become pregnant while taking one. These medications should not be started during pregnancy. Similarly the routine use of SSRI antidepressants is not advised during pregnancy and risk–benefit should be closely assessed in each individual woman.

Management of acid reflux during pregnancy

Over half of women experience acid reflux (GORD or GERD) during pregnancy.[52] This usually starts at the end of the first trimester and continues for the entire pregnancy. It can be really miserable.

Five tips for coping with acid reflux during pregnancy

- **No late-night snacks:** avoid eating and drinking late at night.
- **Tilt the head of the bed:** raise the head of the bed by 10–15cm.
- **Sleep position:** lie on your left side in bed at night (this reduces reflux into the oesophagus).
- **Take simple antacids:** simple antacids, such as calcium and magnesium-based products, are very safe and also give rapid (although sometimes short-lived) relief.
- **Talk to your doctor:** if you're still experiencing symptoms despite all of the above, talk to your doctor regarding a trial of acid-blocking medications to reduce acid reflux. Many of these are tried and tested in pregnancy and have proved to be very safe (such as cimetidine, ranitidine and PPI medications). Always talk to your doctor before starting a new medication (even if over-the-counter) when pregnant.

Conclusion

Heartburn and constipation are common during pregnancy and can usually be managed with a few simple measures. If you have problems with IBS before you become pregnant, then these problems are likely to continue during pregnancy. In general, symptoms are managed very similarly to when you are not pregnant, but obviously greater caution is used when prescribing medications for the first time during pregnancy, and your doctor will always discuss the risks versus the benefits with you. Never take an unprescribed medication during pregnancy without first checking its safety profile with the pharmacist or your doctor.

9

Perimenopause, menopause, your gut and pelvic floor problems

Some women develop significant digestive system problems for the first time around the onset of perimenopause or after menopause has taken place (post-menopause). Other women, who have had problematic IBS throughout their fertile years, may in fact see a marked improvement in their gut symptoms. We are back to the individuality of the human body and the way in which each of us reacts differently. Some women go through menopause with relative ease and may well identify with the idea of 'The Second Spring' – finding pleasure in the benefit of families being reared, financial pressures reduced, careers well-established or even looking towards retirement and new beginnings in a number of areas of life.

For other women, the perimenopause and postmenopause can be incredibly turbulent times on many fronts, and apart from the well-documented effects on mood, concentration, sleep and energy (to name but a few), some women also experience significant changes in digestive functioning. These include:

1. altered frequency of bowel motions – either constipation or diarrhoea;
2. abdominal bloating/distension;
3. problems with pelvic floor function.

Let's look at these in a little more detail.

How do hormone levels affect the gut during and after menopause?

As we saw in the last chapter, oestrogen levels start to fall during our forties and continue to decline during the years of the perimenopause, eventually becoming very low post-menopause. All the effects of oestrogen and progesterone on the gut that we discussed then become less obvious, which is why many women will notice a marked improvement in IBS symptoms during this time. It's not all win–win though, as other symptoms can become more problematic.

1 Altered bowel habit

One of the most common outcomes of the changing hormone levels and the way they impact female digestive function is the development of constipation. Some studies suggest that this may not be due simply to altered sex hormone levels, but also to altered levels of stress hormones like cortisol.[46,54] At this time of your life energy levels may be reduced, mood can be low at times, and you may well be juggling the competing demands of work and family life. The upshot of this is less time and energy for exercise, and reduced exercise can also contribute to slower bowel motility.

In a smaller number of women, the bowel pattern may become more active during or after menopause, such that they develop loose or frequent stools, or even diarrhoea. This may be the result of the physiological changes of menopause, but remember that, at this age, any significant change in bowel function should be investigated and *must* be mentioned to your doctor. There is a particular type of colitis that tends to develop later in life, called 'microscopic colitis'. This is much more common in women and tends to cause very loose, frequent and urgent stools, often associated with incontinence. This is diagnosed by colonoscopy (and biopsies must be taken, as it is not visible to the naked eye during endoscopy).

At this age, any significant change in bowel pattern warrants a colonoscopy.

2 Abdominal bloating

Bloating is a very common digestive symptom, as we have already discussed. Some women, who have never had significant digestive problems, can develop abdominal bloating and distension for the first time, when they begin menopause. There can be a number of reasons for this:

1. **Increased abdominal fat:** sometimes women put on weight around this time, particularly around the waistline, and small degrees of abdominal distension that were previously not too noticeable now become either unsightly or uncomfortable.

2. **Constipation:** if the bowels are not opening as regularly as previously this can contribute to some distension.

3. **Abdominal wall muscle tone:** oestrogen is an important hormone for helping to maintain our muscle mass, and during menopause falling oestrogen levels can lead to a slight weakening of the abdominal wall muscles (along with muscles everywhere else), so that they tend to become a little lax. This allows the abdominal contents to protrude outwards a little, rather than being held in by taut muscles.

4. **Altered GM:** we know that the GM changes during and after the menopause. Various studies have described different changes in gut bacterial subgroups before and after menopause and it is possible that in some people these changes result in bacteria that cause more fermentation within the gut and more gas production.

Can anything be done to help this bloating?
There are a number of things you can do that can help. We'll be talking a lot more about bloating in Chapter 15 too.

Four strategies to reduce bloating around menopause

- **Weight control:** obviously easier said than done! But it is important.
- **Reduce constipation:** measures to improve gut transit include increased fluid intake, increased fibre – possibly fibre supplements – and a judicious cup of strong tea or coffee can be very helpful.
- **Reduce gas production:** a qualified/registered dietitian can devise a personalized diet to help you but you can also help yourself by following a plan like the FLAT Gut Diet, which we'll introduce you to shortly.
- **Exercise:** keep moving! If you can't do abdominal crunches, don't worry; any exercise that uses your core muscles is very beneficial. Pilates and yoga are wonderful in this regard, and many instructors will have classes specially designed for the older woman. Walking at a good pace and swinging your arms gently as you walk will also engage your core. Gentle weights can help maintain muscle mass and this is turn helps control weight as muscle has a higher metabolic rate than fat.

3 Pelvic floor disorders/problems of pelvic floor function

The pelvic floor is the name given to the hammock-like structure of muscles and ligaments that support the pelvic organs. In women, the pelvic organs include, from front to back, the bladder, uterus, vagina and rectum. In men the organs are the bladder, prostate and rectum. The pelvic floor plays a vital role in co-ordinating and controlling urination, moving the bowels and supporting the reproductive organs, and when problems develop in the pelvic floor, the function of any one or all of the pelvic organs can be affected.

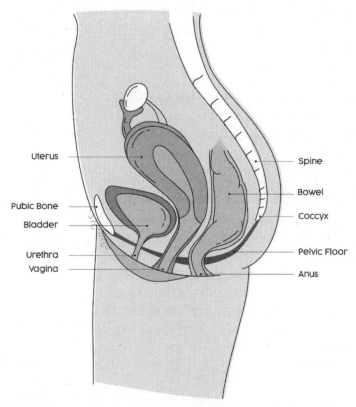

Figure 12 The female pelvic floor: this is a side-view of the pelvic floor, shown as if your body were split down the centre, so the front of the body is on the left and the rear on the right in this diagram

Problems with the pelvic floor are collectively known as pelvic floor dysfunction (PFD). PFD occurs when the pelvic floor becomes weakened or injured and loses its elasticity, which can then affect urinary, bowel or sexual function. Some of these conditions are painful and are sometimes embarrassing for people to discuss, even with their doctor. PFD can have a hugely negative impact on your quality of life and can limit your ability to exercise and engage in normal social activities. Symptoms vary by the type of disorder, and we are going to discuss a few of the more common PFDs, specifically those that affect bowel function, although many of you will also have problems related to control of urination, some pelvic pain or pain during intercourse.

What are the top causes of pelvic floor disorders?

Firstly, women are far more likely than men to get pelvic floor disorders. Additional reasons for this include:

- **Pregnancy and childbirth:** particularly a prolonged or difficult vaginal delivery, as this leads to stretching of the pelvic floor muscles and possibly to anal sphincter damage.
- **Age:** older women are at greater risk as weakening of pelvic muscles and structures occurs after menopause.
- **Obesity:** increased weight, in particular weight around the abdomen, increases pressure on the pelvic floor muscles.
- **Chronic constipation:** straining can lead to stretching of the pelvic floor over time.
- **Repetitive heavy lifting:** women who have done a lot of heavy lifting over their lifetime are at increased risk.

Common symptoms of pelvic floor disorders

- Pelvic pain
- A sensation of 'something coming down' or pressure in the pelvic area
- Difficulty controlling bowel motions/faecal incontinence
- A feeling of a blockage when trying to pass a bowel motion
- A feeling of incompletely emptying the rectum – 'I never feel like I have fully emptied my bowel'
- Urinary incontinence
- Pain during sexual intercourse

Common types of PFD

There are a number of different conditions that can affect pelvic floor function, leading to any one or a combination of the symptoms we have just described. We're going to discuss a few of the more common ones.

Rectocele

A rectocele is a bulge or pouch-like structure in the front wall of the rectum which pushes forwards into the vagina. When a patient with a rectocele strains, the stool may get caught in a pouch and this prevents the patient from emptying the rectum completely. Small rectoceles do not tend to cause symptoms but larger rectoceles may cause difficulty emptying the bowel, or cause leakage of stool after having a bowel movement. Rectoceles are more common in women who have given birth by vaginal delivery. They are caused by misalignment of the sling of muscles supporting the rectum and are often associated with pelvic floor prolapse.

Symptoms of a rectocele:

- A sensation of pressure in the vaginal area
- A sensation of 'something coming down' in the vagina, or you may even feel like you are sitting on a ball
- Difficulty evacuating the bowel motions: you get the urge to pass a bowel motion and you go to the bathroom, but despite the fact that you are pushing, the bowel motions don't seem to come out, or there is a feeling that some of the motion is passed but some remains (this is called incomplete evacuation). If you look at the diagram in Figure 13, you can easily see how the pouch that is bulging forward into the vagina could become filled with stools rather than the stools exiting the rectum through the anal sphincter as they should.

Younger women can also have a pelvic floor prolapse or a rectocele, but for all the reasons that we discussed earlier, they are much more common in older women.

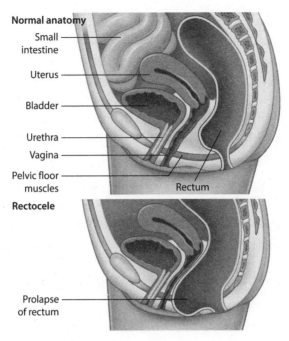

Figure 13 The normal pelvic floor and a rectocele

Pelvic organ/floor prolapse

Sometimes the pelvic floor 'hammock' becomes weakened, particularly after normal vaginal delivery, but also after menopause. This can cause the muscles to sag, and the bladder, womb and rectum can all dip downwards as a result. This is called 'pelvic floor prolapse' and can lead to a very uncomfortable feeling of pressure low down in the pelvis or a feeling of 'something coming down' in the vaginal area.

Obstructed defaecation

Obstructed defaecation is literally a feeling of difficulty passing stools out of the rectum. You may feel that you need to go the bathroom more often, or that you cannot empty completely. Obstructed defaecation may be caused by pelvic floor prolapse, pelvic pain, or a rectocele, among other things. Passing a bowel motion may feel like trying to push against a closed door.

Faecal incontinence

Incontinence of bowel motions means that there is accidental leakage of faeces or wind. Incontinence can be partial, in which a person loses only a small amount of stool, usually when passing gas, or complete, where an entire bowel movement is evacuated uncontrollably. This is one of the most embarrassing things for any adult woman to deal with. People with incontinence are so embarrassed by this condition that they have often been living with it for a long time before they pluck up the courage to mention it to their doctor. They often develop all sorts of mechanisms and routines to try and deal with this dreadful problem, including:

1. Not leaving the house until they have passed a bowel motion.
2. Not going for walks or anywhere far from a bathroom.
3. Immediately identifying all the bathroom facilities in any new location when they arrive (particularly at places like shopping malls).
4. Avoiding going out at all due to fear of being far from bathroom facilities.

Why faecal (and urinary) incontinence is common

- **Labour and childbirth:** natural labour can be traumatic to the muscles of the pelvic floor (they undergo a lot of stretching during labour), such that they never recover their pre-childbirth strength. Prolonged labour, or the need for an episiotomy (cut) can also damage the anal sphincter muscle, which, despite pelvic floor exercises (it is so important to do these), will usually not return to its pre-labour strength. In healthy young women, the muscles are generally quite strong, and they can compensate by squeezing extra hard to keep control of the bowel motions.
- **Low oestrogen:** during and after menopause, low oestrogen levels result in reduced bulk of the muscles in the pelvic area (we lose bulk and strength in all muscles after the menopause) and so muscles that were previously weakened by childbirth are now further weakened by the hormonal changes, and this can lead to an inability to squeeze sufficiently to hold on to bowel motions, particularly if the motions are soft, or a little urgent.

We have met and treated women who are so restricted by these symptoms that they have effectively become agoraphobic as a result. They don't go out to meet their friends, they avoid family occasions or spending precious time with their grandchildren, they stop activities that they enjoy and have retreated into themselves. We do not want women to continue struggling when they don't need to.

What can be done to help PFDs?

PFDs can be hard to diagnose, but if you feel that you have symptoms to suggest any of these conditions and they are causing significant discomfort and/or distress, you should make an appointment to see your GP or/family doctor to discuss them. In our experience, older women can often be very private and find themselves embarrassed talking about digestive functions in general, let alone problems with incontinence. But know that you are not alone: many others just like you have this problem so please don't be embarrassed to speak about it to your doctor. This is a specialized area and referral to a specialist will likely be required; depending on your main symptoms, this could be a gastroenterologist, a colorectal surgeon, a gynaecologist (if the main symptoms seem gynaecological) or a urologist (if the main symptoms relate to problems passing urine).

A number of things can help PFDs:

Five steps to help pelvic floor disorders

- **Diagnosis:** get the correct diagnosis. Talk to your doctor or see a specialist.
- **Pelvic floor physiotherapy:** see a physiotherapist who has a special interest/expertise in pelvic floor conditions. Most of these therapists will be female.
- **Biofeedback therapy:** a specialized form of treatment used by physiotherapists to help re-train the pelvic floor muscles.
- **Posture:** get the right posture for passing a bowel motion – putting your feet up on a step stool can improve the angle of pelvic floor muscles to help passing bowel motions.
- **Medications:** whether you have constipation or diarrhoea, or a combination, sometimes medications can help control the bowel motions, or ease their passage. This can be helpful.
- **Surgery:** this may be an option, or even a necessity, for a minority of people. A specialist would be involved in this decision.

Mary's story

Mary was a fit and active 71-year-old lady when Barbara first met her. She was a retired nurse, mother of four children and proud grandmother of eight. She liked to walk, played bridge and golf, and was very involved with her grandchildren. She was referred to see Barbara by her GP with a complaint of diarrhoea. Like many women of her generation, Mary was quite embarrassed to talk about this problem. It transpired during the course of the conversation that Mary had been having loose stools, approximately two or three times per day since her late 30s. She was young having her children and the first delivery had been difficult – it was a forceps delivery, a large baby boy and she had a second-degree tear at the time. She was 26 and recovered quickly, before going on to have three more relatively 'easy' vaginal deliveries. Over the years she had managed the diarrhoea reasonably well. Without help from any doctor or dietitian, she had discovered that fruit was a problem for her, as were onions and garlic, and she stopped eating these. She also cut down on one of her favourite things – a strong cup of tea. To juggle the demands of family life and work, she had become a practice nurse at a local GP practice, and found that she always had a bathroom close by if needed.

Mary's problem now was that, although the bowel motions had not really changed, she was having some problems with 'holding on to' them – in other words, having episodes of incontinence, or leakage. This was impacting on her ability to play bridge and golf, both things that she loved. This had been going on for some years, but despite the fact that she was a nurse, she was embarrassed to mention it. One of her sons lived in New Zealand, and she and her husband were due to travel there for a six-week holiday a few months after Barbara first saw her. She was very anxious about this because of the problems with incontinence. 'How will I manage at the airport and on the plane?'

On examination, it was clear that Mary had a weakened anal sphincter muscle – she didn't have much ability to squeeze. The rest of her examination was normal. Barbara felt that she had long-standing IBS, but that she also had anal sphincter damage secondary to her previous labours (particularly the first one), now worsened by the age-related post-menopausal changes we discussed earlier. In view of her age, Barbara did some blood tests and a colonoscopy and these were normal. She also had some measurements of her anal sphincter function, which confirmed that the muscles were weak. She was referred to Elaine for dietary assessment and management, and Elaine started her on the FLAT Gut Diet. We also referred her to a pelvic floor physiotherapist to try and to strengthen the anal sphincter muscles. Barbara started her on a bile mop (Chapters 2 and 5), which helped to relieve her symptoms.

Barbara reviewed Mary a couple of weeks before her planned trip to New Zealand. She was doing a lot better, her stools were more formed and down to one or two motions per day. She attended the physiotherapist and, while the results were not miraculous so far, she was doing her exercises every day. While everything had objectively improved and she was feeling more confident about going to bridge and playing golf, she remained very anxious about the trip to New Zealand. We talked things through and agreed that she would take two Loperamide (an anti-diarrhoea medication) tablets before she went to the airport and would take some with her for the journey also. While these might constipate her in the short term, they would give her the reassurance that she was less likely to need to visit the bathroom while travelling and hence less likely to have an 'accident'.

She emailed us when she got back from her trip to say that the journey over and back had been 'uneventful'!

Take-home message Sometimes, a problem with a variety of causes needs to be managed in a number of ways to find a 'solution'. None of the measures undertaken were a cure for Mary's problems, but taken together, they helped make her previously unliveable problems very manageable.

Nature provides! Phytoestrogens

The natural world is truly amazing, and often provides us with what we need. Many plants, including flaxseeds and other seeds, tea, some fruits and in particular soy-based products, are very rich in plant oestrogens, which are called phytoestrogens. Phytoestrogens are converted by the GM into a form of oestrogen called equol,[49] which mimics some of the effects of natural oestrogen produced by the ovaries, although it is not quite as potent. There is some evidence that a diet high in soya-based products may be beneficial to peri-menopausal and post-menopausal women, in helping to reverse the effects of low oestrogen on the body.[55-57] It can reduce weight gain and blood sugar levels, and may explain why there are lower levels of obesity in Asian women who eat a traditional Asian diet, and they live longer than their Western counterparts. Eating more soy-based products is certainly something for peri- and post-menopausal women to think about. One caveat, however – high soya intake is not advised in women who have a history of breast oestrogen receptor-positive cancer.

Take-home messages

- Changes in gut function are common around perimenopause.
- Any significant change in bowel function should be investigated, to ensure that there is nothing more serious going on.
- Pelvic floor problems are common in women during and after menopause – talk to your doctor as a lot can be done to help.
- A number of factors may be contributing to symptoms and it is vital to consider the whole person.
- There have been major advances in the whole area of HRT, and a detailed discussion of this is beyond the scope of this book. You should discuss your options with your GP/family doctor or a specialist in this area.

Section 3

LET'S LOOK AT YOUR DIET

10

What is a healthy diet?

A brief history of what we eat

Over the course of history, people ate food that was in season and which grew in their locality. They were not exposed to food from different cultures and their diets were predictable and limited (and probably very boring). Modern humans evolved between 200,000 and 300,000 years ago. They were hunter-gatherers up to about 12,000 years ago when crop cultivation and animal farming began. Wheat and other cereals started to be cultivated, and gradually over time, have become a staple in many of our diets. These 'Paleolithic' diets of prehistoric humans had advantages over ours: there was little sugar and salt in the diet, and meat was only eaten on occasion. Paleolithic humans ate far more fibre than we do nowadays and almost definitely did not have constipation!

Fast forward to today. We have come to rely more and more on cereals, and on wheat in particular. Wheat consumption has increased over the past 150 years, and not only that, but the type of wheat we eat is different from what our ancestors ate – modern wheat has less protein and more fructans (see Chapter 14) than the older varieties and thus has the potential to cause more bloating. Add to that the fact that we are eating more white bread (with the wheat bran completely removed) and we are clearly eating less fibre. Throw in the effects of globalization and the rise of fast or ultra-processed food (high in sugar, salt and fats) and the upshot of all of this is that we are no longer eating the type of diet that we evolved to eat. We are also a lot less physically active than our ancestors. This is contributing to many of the digestive and other medical problems that we experience nowadays, such as:

- IBS and functional dyspepsia;
- obesity;

- Type 2 diabetes;
- heart disease;
- fatty liver.

What does a healthy diet look like?

Variety is the key to a healthy diet. Nowadays, we have a greater choice of foods than ever before. We're going to take a whistle-stop tour of the nutrients that make up a healthy diet, before moving on to specifically discuss diet and IBS, and introduce you to the FLAT Gut Diet. This may bring you back to your school biology classes.

The three main components of a healthy diet	
Component	Function
Water and any other fluids you drink	To hydrate the body: vital for the functioning of every cell in the body and, along with fibre, aids gut motility.
Macronutrients	These are the main 'building blocks' of the food we eat and are the source of energy (calories). Protein Carbohydrate Fat
Micronutrients, including minerals and vitamins	These are vital substances for the function of cells in our body. Our body cannot make these and so we *must* get them from our diet. They do not provide any energy (calories).

Let's begin with fluids and what are called 'macronutrients' – protein, carbohydrates and fats. We'll then take an overview of micronutrients, particularly those that play a significant role in female health.

Fluid

Water is a vital component of our daily diets, and if you do not drink enough fluid you can become dehydrated, which in turn can cause you to feel lethargic and to develop headaches. Your daily water requirement is based on your body weight – smaller people need less water. A good guide is 30–35ml of water per kg body weight. So if you weight 60kg, this works out at about 1.8–2L per day. In the USA, the recommendation is a half to one ounce of water per pound body weight. You can actually have too much of a good thing – too much water can lower the salt level (sodium) in your blood and cause dizzy spells, light-headedness and even seizures.

Hydration tip

A good way to know if you are drinking enough fluid is to look at your urine. First thing in the morning urine is always dark yellow. If you are well hydrated you will pass urine (pee) four or five times during the day and the urine will be very pale yellow. If you're not drinking enough you'll pass much less urine and it will be darker in colour.

Macronutrients

Protein

Protein foods provide us with amino acids, which are the building blocks of every cell in our body. There are 20 amino acids; we can make 11 of these ourselves in our bodies but nine can only be obtained from our diet. These nine amino acids are called essential amino acids and we *must* get them from our diet.

Animal-based proteins are known as complete proteins as they contain all essential amino acids. These include meat, poultry, fish, eggs and dairy foods. A few plant-based foods are also complete proteins such as buckwheat, chia seeds, hemp, and quinoa. If you eat a plant-based diet it is extremely important to ensure that you get these essential amino acids in your diet, or if not, that you take supplements. Many other plant-based proteins are called 'incomplete' as they do not contain all essential amino acids. These include grains, legumes, nuts, seeds and some vegetables. Protein foods supply plenty of other beneficial nutrients, including iron from red meats, healthy fish oils, calcium for bones and many vitamins and minerals.

You should be having two servings of complete protein per day and three servings of dairy or calcium-fortified plant-based protein to meet daily calcium requirements (see the 'Calcium requirements' table, below).

> ### Protein tip
>
> Protein foods keep you fuller for longer, repair and maintain muscle, suppress the appetite and help you maintain a healthy weight.

Carbohydrate

Carbohydrates are basically made up of simple sugars or chains of sugars, and they are a major source of energy (calories) for our bodies. The chains of sugars are called 'complex carbohydrates' and don't cause a 'sugar rush' after you eat. Instead, the sugars are slowly released from the food and into the bloodstream, in a much more controlled way. You might have heard of the glycaemic index (GI) of food – this relates to how quickly sugars are released from a given food. Simple sugars like glucose (in sweet foods) are rapidly absorbed and have what is called a high GI (this is not good), whereas the 'complex' carbohydrates have a low GI (good).

The main sources of carbohydrate are:

- grains and cereals of any kind;
- anything sweet;
- fruit and vegetables.

Complex carbohydrates are high in fibre, low in fat, very filling, feed your GM and contain important vitamins and minerals. These should be your choice of carbohydrate to optimize your energy, gut health and general wellbeing.

Simple carbohydrates include many 'white' foods including white cereals, breads, crackers, flour, sugar, biscuits, cakes, confectionery, juices and sweetened drinks. These are basically 'empty' calories and too much of these foods is linked to the current obesity epidemic, tooth decay, heart disease, cancers and diabetes.

The recommended carbohydrate servings are from three to five per day but are very much dependent on age, gender, body size and activity levels. If you have IBS or functional dyspepsia, the choice of carbohydrate foods is extremely important in helping to reduce your symptoms. We will tell you exactly how to manage your carbohydrate and fibre intake in the FLAT Gut Diet in Section 5.

Carbohydrate tip

Weight loss and stubborn abdominal weight can be aided by reducing your daily carbohydrate intake. Not NO carbs, but smaller portions mainly at your main meal of the day.

Vegetables, salad and fruits

These are carbohydrate foods, but are not thought of as 'carbs', when we talk about 'cutting down on carbs'. They contain complex carbohydrates, and apart from some fruits that have a high amount of fructose, they generally do not contain much simple sugar. They are an excellent form of fibre, vitamins and minerals, and they are low in both calories and fat.

Ideally you should 'eat the rainbow', which means that you should choose a wide variety of coloured salads, vegetables and fruits. Variety is key in any healthy diet as it ensures that you get your fibre from different sources, and that you enjoy the benefits of the many different 'phytochemicals' (naturally occurring chemicals present in plant-based foods). So whenever you include a healthy portion of vegetables with your meal, or savour a handful of juicy berries, not only are you getting the benefits of the minerals, vitamins and fibre in these foods, but you are also getting an extra bonus via the various phytochemicals they contain, many of which have very particular benefits. These include anti-inflammatory effects (reducing autoimmune tendencies), immune boosting, cholesterol lowering, weight control, anti-cancer activity, to name but a few. We should aim for two to three fruit portions per day and four to five salad and vegetable portions daily. In the table below, we show just a few of the phytochemicals known to be of benefit to health.

Phytochemical	Food source	Health benefit
Phytosterols	Nuts, seeds and fruit	Anti-cancer
Polyphenols	Fruits, vegetables, cereals, chocolate, legumes, oilseeds	Antioxidants, protects against inflammation and cell damage
Phytoestrogen	Legumes (particularly soybeans), berries, red grapes, peanuts	Can mimic the effects of oestrogen and help reduce the effects of menopause, and protects against bone loss
Glucosinolates	Cruciferous vegetables (the cabbage family, bok choy, broccoli)	Protects against colon and stomach cancer
Saponins	Oats, green tomatoes	Protects against infection
Terpenoids	Mushrooms	Protects against infection
Gingerol	Ginger	Antioxidant, anti-inflammatory, may reduce nausea, may help to reduce weight
Curcumin	Turmeric	Antioxidant, anti-inflammatory, protects against cancer, may reduce depression, reduce the risk of heart disease and Alzheimer's

Fibre, veg, salad and fruit tip

These foods prevent disease, promote vibrant health, are very filling, aid weight control and are packed with minerals, vitamins and phytochemicals.

Fat

Fat has been demonized for years and was blamed for the 'obesity epidemic'. We now know that this is not true and that high intakes of 'simple' carbohydrates and sugar are in fact the culprits. Foods containing fat provide energy and essential fat-soluble vitamins (A, D, E, K). It is not necessary to consume low fat products and, in fact, we should include more healthy fats in our diets. Animal fats (saturated) are less healthy than fats from fish and plant-based sources such as vegetable oils, nuts, seeds and avocado: these are what are called mono- and poly-unsaturated fats. This is the basis of the Mediterranean diet.

> **Fat tip**
>
> Increase your intake of oily fish and vegetable oils to benefit from 'good' fats.

Now let's take a look at all the important micronutrients for a woman's health.

Micronutrients

Iron is an essential mineral, vital for the formation of red blood cells, which carry oxygen around the body. Low iron levels tend to cause tiredness and fatigue. Menstruating women and pregnant women are particularly prone to iron deficiency. In general, animal-based foods are a richer source of iron than plant-based foods, and the iron is also more easily absorbed from animal sources. For these reasons, it can be challenging to meet your iron requirements if you follow a plant-based diet, particularly if you are vegan. Vitamin C helps absorb iron.

Calcium is essential in the body for the formation of bones and teeth. Calcium is found in lots of foods, but as with iron, the calcium content of plant-based sources is poor. We have included an easy to follow chart to allow you to calculate your daily calcium intake. If you are not meeting your daily calcium requirements, you should take a calcium supplement.

Calcium requirements

Group	Age (years)	Calcium (mg) per day
Adolescents	11–18 years (girls)	800
Adults	19+ years	700
Breastfeeding mums		1,250
Women post-menopause		1,200
Coeliac disease/ inflammatory bowel disease	Adult	1,000
Osteoporosis	Adult	1,000

(Continued)

Calcium content of suitable foods		
Milk	**Portion size**	**Calcium content (mg) per portion**
Milk	100ml	120
Milk super fortified	100ml	160
Lactose-free milk fortified	180ml glass	210
Almond, hemp, nut, quinoa, rice milk*	180ml glass	216
Yogurts		
Whole milk/low fat/ drinking yogurt	100g	170
Cheese		
Cheese (naturally low in lactose)	30g (matchbox size)	220
Cottage, cream	150g	190
Fish		
Tinned sardines with bones	2 fish (50g)	260
Tinned salmon	100g	93
Plant-based alternatives		
Tofu	50g	250
Brazil nuts	30g	51
Vegetables & fruit		
Spinach	100g	99
Oranges	1 medium	75

*Calcium content of fortified milks varies; it is very important to check the calcium content of each individual product.

Vitamin D, often referred to as the sunshine vitamin, is vital to enable you to absorb calcium. The majority of vitamin D is synthesized in your skin when you are exposed to ultraviolet light from the sun. Vitamin D can be found in oily fish and, in small amounts, in beef, liver, cheese and in fortified breakfast cereals, milks and spreads. Most of us don't get enough vitamin D in our diet. It's extremely important to take a vitamin D supplement, particularly during the winter months, or if you wear sunscreen during the summer months. People who live in sunny climates tend to have higher vitamin D levels than those who live in colder parts of the world. Women who for religious reasons wear head-coverings, or whose skin is not exposed to the sun, are also at high risk of vitamin D deficiency. This is particularly the case for Muslim women, and they should strongly consider taking a vitamin D supplement for bone health.

Magnesium is another essential mineral for bone health and is often overlooked. It contributes to increased bone density and helps prevent the onset of osteoporosis. Magnesium is found in nuts, legumes, tofu, seeds, whole-grains, leafy green vegetables and some oily fish, so if you follow a plant-based diet, you will generally have plenty of magnesium.

Vitamin B12 is a very important nutrient for every cell in the body, but in particular for function of our neurons and for production of red blood cells. If you are low in B12 you may feel fatigued, become anaemic or develop pins and needles. The main sources of vitamin B12 are meat, fish, chicken, eggs and dairy foods, so if you follow a plant-based diet, you need to ensure that you get enough B12 either through supplements or through foods, such as breakfast cereals and plant-based milks, which have been fortified. Fermented soya products and tofu are good sources of vitamin B12, but the B12 is not that well absorbed from these foods.

Folate, or **folic acid** as it is also called, is vital in a woman's diet if trying to conceive to prevent neural tube defects, such as spina bifida, in foetuses. A daily supplement of 400ug folic acid daily is advised for any woman who is hoping to conceive. Folic acid is found in green vegetables and, if you are restricting your vegetables to control IBS symptoms, it is important to check your folate levels.

Food sources of important dietary micronutrients
Omega 3 – oily fish (such as anchovies, herring, mackerel, salmon, sardines, trout, tuna) and avocado Nuts and seeds – chia and flaxseed seeds, walnuts Plant oils – flaxseed, canola, cod liver, rapeseed, soya bean
Iron – red meats, liver, eggs, black pudding Fortified breakfast cereals, legumes, nuts, seeds, soya and green leafy vegetables
Calcium – milk and dairy foods (milk, yoghurt, cheese) Fortified plant, nut, oat, soya, rice milks and cereals Fish – with bones (such as salmon and sardines) Tofu, beans (baked, kidney), chickpeas Green veg – broccoli, kale, okra, spinach, spring greens Nuts, fruit, seeds – almond, brazil, hazelnuts, apricots, figs, oranges, sesame seeds
Vitamin D – fortified cereals, milks and yogurts Oily fish – mackerel, salmon, sardines, trout, tuna Eggs and sunshine
Magnesium – nuts, seeds, peanut butter, legumes, tofu, seeds, wholegrains. leafy green vegetables and some oily fish
Vitamin B12 – meat, dairy foods, eggs, Marmite, nutritional yeasts, fortified breakfast cereals and fortified plant-based milks, fermented soya products, tofu
Folate – fortified cereals, asparagus, broccoli, kale, spinach, baked beans, orange juice
Phosphorus – meat, poultry, dairy products, fish, beans, nuts and seeds
Selenium – Brazil nuts (best source), crab, lemon sole, mussel, sunflower seeds
Iodine – fish, shellfish dairy foods, eggs, kelp, seaweed
Zinc – legumes, nuts, seeds and soya products

Nutrient deficiencies in IBS

As we mentioned earlier, studies have shown that up to 15 per cent of people with IBS restrict their diet so much that they become deficient in some important vitamins.[35,58] Add to that the fact that many women's diets are already quite low in fibre, iron, calcium, folate, vitamin D and magnesium and you can see that this really is a recipe for significant nutritional deficiencies. These deficiencies can creep up on people, unrecognized over many years. We have treated young women (under 50) who have developed osteoporosis due to diets low in vitamin D and calcium. They find out about the osteoporosis when they sustain a fracture after a seemingly minor injury. Don't let this be you. Don't follow fads. Make informed decisions about what foods to eat and what you are going to avoid.

Annual Nutritional Blood Panel for Women	
Full blood count (FBC)	Serum iron
Serum ferritin	Vitamin B12
Folic acid	Calcium
Vitamin D	Phosphate
Magnesium	Fasting lipid profile (for over 50s)

Mediterranean diet

A healthy diet contains a good blend of all the macronutrients and micronutrients you need, along with plenty of fibre and plant-derived phytochemicals. While more and more people nowadays are choosing to follow exclusively plant-based diets, because of the many benefits of plant-based foods, we believe that a Mediterranean style diet ticks all the boxes when it comes to gut and total body health. This diet allows you get essential amino acids from animal- and fish-based sources and yet is high in good fats from fish, olive oil, nuts and seeds and generally low in animal fat. It is packed with all the essential vitamins and minerals and includes a wide range of plant-based foods from fresh fruits, grains and vegetables, all of which promote diversity of our GM and optimize total body health. More importantly for the patients we see and for people with digestive problems, a Mediterranean-style diet is easier to tolerate and causes less bloating than an exclusively plant-based diet.

Our FLAT Gut Diet, developed for the management of IBS, is based on a healthy Mediterranean-style diet, which we have modified to ensure that the foods chosen minimize gut symptoms in people who have IBS. We are excited for you to get started on this diet so that you can get your gut symptoms under control, while enjoying the many benefits of healthy eating.

Take-home messages

- **Water:** make sure that you are drinking enough, but not too much.
- **Protein:** keeps you full for longer and tends to be very well tolerated if you have IBS.
- **Plant-based foods and fibre:** these have many benefits for your overall wellbeing and gut health and should make up a large part of your diet.
- **Good fats:** fats derived from fish and plant-based sources are generally very healthy and are a rich source of fat-soluble vitamins, in particular vitamin D, which is vital for your bone health.
- **Vitamins and minerals:** women can easily become deficient in a number of micronutrients and it's good to be aware of their sources in your diet and to have blood levels checked.

11

Why we love fibre: super-food for your gut

Fibre is a term used to describe a range of plant-based carbohydrates, which, unlike other carbohydrates, are not digested in the small intestine and so reach the colon in all their undigested glory. Once in the colon, the fibre acts as a rich food source for the GM, stimulating growth and proliferation of the bacteria and also the production of a number of beneficial substances, which are vital for the health of the gut. We are passionate about the health and gut benefits of fibre and we're delighted that fibre has been making a comeback recently.

Fibre adds bulk to the stools and this stimulates gut motility. It also holds on to water in the stools, keeping them soft and lubricated, all of which helps the passage of a bowel motion and prevents constipation.

What is fibre made of?

Fibre comes from plant-based foods, where it is present in the cell walls of the plants and within the cells (this is called resistant starch). There many different types of fibre and they are all composed of various combinations of chains of simple sugars of varying lengths, some with lots of side chains also. Humans do not have the digestive enzymes (glycoside hydrolases) required to break down most of these chains of sugars and so they pass through the gut, undigested, until they reach the colon, where many are digested by the GM (the gut bacteria have the enzymes required to break down the sugar chains). The GM in turn produce important substances like short-chain fatty acids (SCFAs) which are very important to the health of the cells lining the gut and are their main source of energy.

Fibres made of fructose chains with glucose at one end are called fructans; shorter chains are called fructo-oligosaccharides (FOS) and the longer chains are called inulins. Galacto-oligosaccharides (GOS) are chains of lactose sugars, found mainly in legumes. This may seem like a lot of detail, but the reason we are mentioning all of these terms is that if you have IBS, and are trying to empower yourself with information, you will likely across these terms as these are 'FODMAPs'.

Many fibres are also classed as 'prebiotic' foods, which means that they stimulate growth of the GM and have a proven health benefit. However, apart from those fibres that are classed as prebiotics, the health benefits of fibre in general in the diet have been proven in many studies.[9] In fact, fibre is one of the best-studied food substrates in the treatment of IBS.

Fibre has been shown to have many benefits for your overall health, as well as your gut health, and here are just a few of these:

Overall health benefits of fibre in the diet

- Reduces risk of bowel cancer
- Reduces risk of breast cancer in women
- Reduces obesity
- Reduces the risk of heart disease and stroke
- Reduces the risk of developing Type 2 diabetes
- Reduces cholesterol levels
- Reduces death from all causes

Benefits of fibre for the gut

- A diet high in fibre helps to resolve and prevent constipation.
- Fibre is necessary to promote peristalsis, which moves food through the colon.
- High fibre foods absorb water and expand the inside walls of the colon, easing the passage of waste.
- As fibre passes through the intestine undigested, it absorbs large amounts of water, resulting in softer and bulkier stool.
- Soluble fibre is beneficial in IBS.
- Fibre provides a rich food source for our friendly GM.
- Fibre stimulates the GM to produce short-chain fatty acids (SCFAs), which provide energy for the cells lining the gut and are vital to the health of the gut.

If there were a pill that did all of this, everyone would take one! Yet more and more people do not have adequate fibre in their diets. The standard Western diet is low in fibre and high in fat, whereas the Mediterranean diet, of which we are big supporters, contains plenty of fibre, along with lots of other beneficial nutrients.

All fibres are *not* the same

To think of all fibres as being the same is a bit like going to your favourite department store and saying that all the clothes are just, well, clothes – and ignoring all the different styles, colours, sizes and uses of all the different clothes available. The same thing applies to fibres; they vary in many ways and certain fibres act as food for different bacteria within our GM. Some bacteria will feed on fibre that comes from grains, for example, whereas others will prefer to feast on fibre that comes from certain fruits or legumes. Fibres can be classified in a number of ways; the old-fashioned classification described fibre as soluble and insoluble. Now we describe other characteristics of fibre as well. There are long- and short-chain fibres, viscous (gel-forming) or non-viscous fibres and finally there are fermentable and non-fermentable fibres. Fermentable fibres are devoured by the GM and lead to gas production, whereas non-fermentable, or low-fermentable fibres do not.

Our digestive system thrives on having a mix of different fibre types in the diet, as each type functions in a slightly different way, providing food for different gut bacteria in the GM, and all support the healthy functioning of our gut. In general plant-based foods that appear rough in texture, or have a hard skin, or stringy flesh, are high in insoluble fibre. These tend to be less fermentable. The key is *diversity* – you get lots of different fibres by eating lots of different plant-based foods. This is why you are being told to 'Eat the rainbow' – lots of colour from different plant-based foods ensures that you are getting fibre from diverse sources. What's more, most fibre-rich foods contain a mix of fibre types.

These features of different fibre types are very important in people with IBS, where excess gas production can result in abdominal bloating, distension and pain. In the FLAT Gut Diet, we will help you to eat plenty and varied fibres, which are less fermentable, meaning that they tend to cause less bloating and flatulence, both of which are helpful if you have IBS or other gut problems.

Effects and sources of fibre	
Effects on the gut	
Soluble fibre	Insoluble fibre
Forms a gel and holds water in the gut.	Not soluble in water, but does attract water into the bowel.
Thickens and lubricates the stools.	Stimulates gut transit and can help constipation.
Fermented by the GM – produces SCFAs but also produces gas as a by-product.	Adds bulk to the stools.
Helpful in IBS-C and also in IBS-D, as the gel-like stools can also slow down the gut transit in those with diarrhoea.	Less fermented by the GM than soluble fibre.
Longer chain fibres, such as psyllium, tend to be less fermentable and more slowly digested by the GM and produce less gas.	Can worsen IBS symptoms possibly due to bulking effect.

Sources	
Soluble fibre	Insoluble fibre
Oats	Wheat bran
Oat bran	Wheat cereals
Barley	Brown rice
Rye	Seeds
Psyllium	Vegetable & fruit skins
Nuts	Whole grains
Linseeds	Corn
Citrus fruits	Cabbage family
Fruits	Legumes
Berries	Nuts
Root vegetables	
Legumes	

How much fibre should we eat?

There is no precise figure for exactly how much fibre we should eat in a day, as evidenced by the fact that the food and health authorities in different countries have different daily fibre recommendations; below we show recommended figures for the UK, USA and our own country, Ireland. Many people are eating far less than the recommended amounts, sometimes as little as 5–10g per day. As with everything, people differ in their response to fibre, but a general fibre recommendation ranges from 21–35g per day. We recommend a daily fibre intake of 20g as a starting point, and to increase gradually from there, according to your individual tolerance. If you have not been eating much fibre in your diet for some time, then it is important to start to increase your fibre intake gradually or you are likely to develop unpleasant side effects of bloating, abdominal pain and excessive flatulence. Try increasing your fibre intake by about 5g per week (start off low and go slow).

Recommended daily dietary fibre allowance	
UK	30g – increased from 18g in 2015 (general adult recommendation)
USA	21g (female, over 50), 25g (female under 50)
Ireland	24–35g (general adult recommendation)

Is there a downside to fibre in the diet?

The short answer is yes:

- People with IBS are more sensitive to the effects of fibre. Fibre increases stool bulk, which distends the colon and this can increase abdominal pain and bloating. In some women too much soluble fibre can worsen symptoms of constipation by forming too much of a gel like substance, which can actually slow bowel transit.
- Equally, with insoluble fibre, particularly wheat-based fibres, many patients experience bloating and excess flatulence. Wheat bran has been reported to worsen symptoms in over half of patients with IBS, while only ten per cent reported an improvement in constipation.[59]

- A high fibre diet may also cause an increase in bowel frequency and diarrhoea.
- There has been a lot of health promotion recommending us to increase our daily fibre intake from fruit, vegetables and salad. Many health-conscious people, who have aimed to meet these guidelines, have in fact exceeded the recommendations, and this can lead to IBS-type symptoms, or to worsening of IBS symptoms.
- It's all about balancing enough fibre to allow bowel transit, while minimizing the side effects. You can do this by choosing a variety of fibre-containing foods, and focusing on your own tolerance level to avoid making your symptoms worse. There is a detailed chart with the fibre content of many different foods in Chapter 20, and we will explain how to find the balance of fibre that is right for you.

Pippa's story

Pippa had a consultation with Elaine as she was diagnosed with IBS by her family doctor and was complaining of loose stools and abdominal bloating. Elaine took a detailed dietary history from Pippa, and the chart below shows a sample day of Pippa's diet.

Pippa's diet is an example of how you can have too high a daily intake of fibre.

Meal	Food	Fibre content
Breakfast	Bran flakes with 10 raspberries & 2 tsp flaxseed/linseed	11g
Snack	Spinach & kale smoothie (with chia seeds)	15g
Lunch	Large salad (lettuce, tomato, cucumber, spring onion, radish, pepper, sunflower seeds)	9g
	2 slices wholemeal bread	4g
	Protein source (60g chicken, beef, turkey, pork, lamb, fish, egg, cheese)	–
Snack	One orange	2g
Evening meal	3 x portions of vegetables	6g
	Brown rice (100g cooked weight)	2g
	Protein source (chicken, beef, turkey, pork, lamb, fish, egg)	–
	Total	49g

Pippa was consuming too much fibre for her tolerance level. The high fibre dietary content was causing IBS-type symptoms, and in her case, her symptoms were easily controlled by a reduction in her fibre intake.

Take-home message We all need to be aware of our daily fibre intake and to find our own individual tolerance level. When it comes to gut symptoms, too much fibre can be as bad, if not worse, than too little.

Conclusion

We are back to the Goldilocks scenario! Fibre is incredibly important for our gut and general health and the long-term health benefits are beyond doubt. Diversity of fibre sources is key, and you get this by 'eating the rainbow'. However, if you have IBS, it is absolutely vital that you find your daily tolerance level for the amount and type of fibre, and that you get this 'just right'! We'll show you how to do this in the Simplify and Challenge steps of the FLAT Gut Diet.

12

Is it what you're eating? Do you have a food allergy or a food intolerance?

Most of our patients feel that their gut symptoms are triggered by food, either by the very fact of eating anything at all, or eating certain foods.[35] People wonder if they might be 'allergic' to certain foods. We mentioned before that patients with IBS often exclude more and more foods from their diet, in search of a solution to their problems, and unwittingly risk developing nutritional deficiencies as a result.

People are getting 'advice' from so many different directions these days – family, friends, influencers, the media, Google – it's not surprising that they are confused, frustrated and worried about their diets. Never have we had so much nutritional information available and yet, confusion reigns. If all this dietary information was helping people, then why are we are seeing ever more patients at our clinics with IBS-type symptoms and functional dyspepsia? We want you to have the information to make evidence-based food choices, which will help your overall and digestive health and minimize your IBS-related gut symptoms.

Dietary Confusion Reigns?

It is not always easy to pin down if a certain food is triggering symptoms, but we will start with some basic principles. One of our passions is making evidence-based decisions before altering consumption of any important food group in the diet, such as dairy (rich in calcium and often fortified with vitamin D) or wheat (a great source of fibre). Let's start with the difference between food allergies and food intolerances. Food intolerance is much more common than food allergy and plays a large role in IBS, but we will briefly discuss some common food allergies, before talking in more detail about food intolerance.

Food allergy

True food allergy affects only around one to two per cent of adults and four to six per cent of children (many children grow out of their allergies). Allergies occur when the body's immune system gets confused and thinks that the protein content of certain foods is harmful.[60] As a result, the immune system is activated when we are exposed to the particular food, leading to allergic-type symptoms. Allergic symptoms can be immediate or delayed. If the response is immediate, it is usually easy to identify the culprit food (such as severe nut allergy) but it can be much more difficult if the reaction is delayed. Mild symptoms of an allergic reaction include itching, hives, skin rash, swelling of the tongue or lips, runny nose, vomiting, upset stomach, bloating, nausea or diarrhoea. At the more severe end of the spectrum allergies can cause wheezing, breathing difficulties and anaphylaxis (life-threatening circulatory system collapse).

The majority of allergic reactions among children are due to egg, peanut, cow's milk, fish, shellfish and various nuts. The most common food allergy in adults is a relatively unknown condition called pollen-food syndrome and this accounts for around half of allergic reactions among adults – and yet you have probably never heard of it.

Common food allergies

Pollen-food syndrome (PFS) also called pollen-food allergy syndrome (PFAS)

Also known as oral allergy syndrome, or fruit and vegetable allergy,[61] it affects about two per cent of adults in the UK and USA. It is more common in people who have latex allergy and hay fever (allergic rhinitis), and in fact is very uncommon in people who do not have hay fever. It is caused by allergy to cross-reacting proteins found in pollen, raw fruits and vegetables and some nuts. People can usually tolerate the culprit food in cooked form, as the proteins are altered during cooking and no longer trigger an allergic response. The symptoms of PFS are usually quite mild and occur immediately after eating the food. They include mild swelling of the tongue and lips or some tingling or itching around the mouth, and the symptoms tend to settle down very quickly. PFS does not usually cause other gut symptoms.

The most common food triggers are raw apples, kiwi, brazil nuts, walnuts, hazelnuts, almonds and cherries, but can involve many more fruits and vegetables.

Latex-food allergy syndrome

This occurs in up to half of people who have latex allergy, again due to cross-reacting antibodies. Patients have a hypersensitivity to a number of different fruits and vegetables. Banana, kiwi, avocado and chestnuts are the most common food allergies, but some people will find many different fruits and vegetables a problem, including potatoes, tomatoes and celery. Some people with latex-food allergy can get symptoms when peeling raw potatoes, but can eat cooked potatoes without any problems. The symptoms of latex-food allergy are similar to those of PFS, and tend to be mild, although on rare occasions they can be severe.

Cow's milk protein allergy

Cow's milk allergy is caused by an allergic response to milk protein (as opposed to lactose intolerance, of which more later). Cow's milk allergy affects about one per cent of children and half a per cent of adults. Treatment is complete avoidance of cow's milk and all related products. The symptoms of cow's milk allergy vary considerably – they can be mild or severe. What's more, they tend to be delayed, and so it can be difficult to make a link between the milk you had with your breakfast cereal and the symptoms you get mid-afternoon or even the following day. This makes it one of the most difficult food allergies to diagnose. The symptoms can be very similar to those of IBS.

Michelle's story

Michelle, 52, was referred to Elaine by a gastroenterology colleague, following normal standard investigations with a diagnosis of likely diarrhoea-predominant IBS (IBS-D). She had life-long symptoms of diarrhoea, three to six motions per day associated with urgency and excess wind. Sometimes she just made it to the bathroom in time. This was having a negative impact on her daily life as she travelled with work, and the fear of having an 'accident' was causing her stress and anxiety.

Initially Elaine felt that the symptoms were indeed consistent with IBS-D and started Michelle on the FLAT Gut Diet. Her response was disappointing, and though she felt that she was 40–50 per cent improved, the stools were still loose. Delving into things a little further, Elaine established that there was a strong history of allergy in Michelle's family. Michelle and her father were mildly asthmatic and a sister had hay fever. She also vaguely remembered as a child something about milk not agreeing with her. A dietary history showed a very large intake of milk and dairy products. Michelle was consuming 0.5L of fortified milk per day, in addition to cheese and yoghurt, as her mother had osteoporosis and she was afraid of developing it too.

Elaine suspected a cow's milk protein allergy for all the above reasons and also the lack of expected response to the FLAT Gut Diet.

Cow's milk protein allergy was confirmed on skin prick test by a consultant allergist.

Elaine prescribed a completely cow's milk protein-free diet, with calcium and vitamin D supplements. Within days of starting the diet, Michelle's symptoms had almost completely disappeared.

Take-home message Keep an open mind; IBS symptoms can be mimicked by other conditions, so if an IBS-targeted diet does not help, think again. Everything in your medical history is important. Food allergies are much more common in people who have other allergic conditions such as asthma and hay fever.

Other commonly recognized food allergies*

- Egg
- Peanut and nut
- Fish and shellfish
- Wheat protein (non-coeliac)

- Soya
- Celery (celeriac-root)
- Sesame seed
- Mustard seed

*Most of these allergies are immediate-type, can be severe and necessitate someone carrying an adrenaline pen.

Diagnosing food allergies

Most immediate-type allergies are easy to recognize and are associated with IgE-type antibodies to the culprit food. Immediate-type reactions cause the 'typical' anaphylactic response that people know about: wheezing, swelling of the mouth, throat and tongue, severe breathing difficulties and even death if not treated immediately (with an adrenaline pen).

Many more delayed-type allergic reactions are not mediated via antibodies and very often have a negative IgE antibody to the particular food, which makes diagnosis more difficult. In this situation a careful dietary and medical history must be taken, ideally by a consultant immunologist or allergist, followed by validated allergy testing, such as skin prick testing and strict food elimination diets. This can be challenging and time-consuming, but very worthwhile if the culprit food is found.

Validated food allergy testing

1. Blood IgE levels
   A blood sample is taken (not a finger prick sample) to measure levels of IgE for foods in the blood. IgE are the type of antibodies in your blood that can be involved in allergic reactions. The most common foods tested are IgE levels for milk, wheat, fish, egg, peanut and soya. If the results are outside the range provided this *may* indicate a potential food allergy. Tests can be 'falsely positive', so if levels are elevated a skin prick test may be recommended to ascertain severity of the allergy. Similarly a negative test does not rule out a specific food allergy, as many food allergies are not IgE-mediated.
2. Skin prick test
   In this test, the suspected food allergen is placed on the skin, in a skin prick test. If the individual has a allergy to the particular

food within about 10–15 mins they develop a red raised bump, indicating an allergy. This test is normally carried out in a specialist allergy clinic or by a consultant allergist or immunologist and does require interpretation.

Unvalidated food allergy and intolerance testing

There are a growing number of different tests available, which claim to test for food allergy or intolerance. People with chronic gut problems, like IBS, are often frustrated and desperate to try to find a cause and a solution to their problem, and will often try anything that might help them.[62] This includes trying some of these commercial tests, which are usually very expensive. Apart from the cost, our major concern with these tests is that they often lead people to overly restrict their diets, which, as we mentioned before, can lead to significant nutritional deficiencies.

IgG-based food allergy and intolerance tests: are they of value?

The short answer to this question is 'no'. If you are interested in knowing a little more, read on.

This is a very controversial and sometimes emotive topic. There are many commercially available food allergy 'finger prick' tests in use. These tests give read-outs for IgG levels to a vast array of food proteins. IgG antibodies are not involved in allergic-type reactions and are produced by our body in response to exposure to foreign proteins. A subgroup of IgG called IgG4 is also used in some commercial tests, on the basis that IgG4 is involved in triggering histamine release, however there is no evidence to support this in human food allergy.[63] Based on the readings obtained from these commercial tests, people are then advised to eliminate foods for which they have high IgG levels. The problem with this approach is that it is entirely normal to have IgG antibodies to foods that we eat and many experts believe this indicates exposure and *tolerance* to the foods, not allergy. For example, if you eat a lot of bread or garlic or avocadoes, or beef for that matter, you will have higher IgG antibodies against these foods. If you stop eating these foods for a period of time, your antibody levels will fall.

Several studies have looked at the benefit of using IgG-based testing as a basis for dietary changes in patients with IBS and the results are inconclusive. One early study found some benefit,[64] but there were significant criticisms of that study. Other studies have shown no differences in food IgG levels between patients with IBS and healthy controls.[65]

Based on the current evidence, the American Society for Clinical Chemistry (AACC), the Canadian Society and The European Society all state categorically that IgG testing for food intolerance in invalid and should not be used.[63] We agree! These tests are very expensive and have no clinical value.

Other non-validated tests

A number of other tests are currently in use, particularly in alternative and complementary medicine practices. These include iridology (studying the iris of your eye), hair analysis, electrodermal testing, cytotoxic assays, stool pH (acidity) testing, kinesiology, to name but a few. We do not recommend the use of any of these tests.

Claire's story

Claire was 38 years old when she self-referred to Elaine for a dietetic consultation. She had long standing symptoms of bloating (extreme at times), excessive wind, borborygmi (noises in the abdomen) and intermittent loose bowels. She described fatigue associated with the extreme bloating. She described how these gut symptoms were interfering with her daily life as a busy working mother of three young children.

Claire was a non-smoker and had a minimal alcohol intake. She was fit and healthy on presentation, at a healthy weight of 64kg (140lb) and a normal BMI of 24. She exercised, mainly by brisk walking, three to four times per week. She had no family medical history of allergies or any other digestive conditions. Her mother was diagnosed with osteoporosis in her 60s.

Claire told Elaine that her digestive symptoms had been present since her early 20s. Recent bloods carried out by her GP were all normal, including coeliac tests. Claire had just recently been diagnosed with osteoporosis after she sustained a nasty wrist fracture following a very minor fall. A DEXA scan had been recommended which confirmed osteoporosis. She had just been started on vitamin D/calcium supplements and Prolia (Denosumab) injections for osteoporosis.

At the age of 29 Claire saw a nutritional therapist (not a dietitian or clinical nutritionist) who carried out a non-validated IgG test as we discussed above. The findings falsely indicated sensitivity to wheat and dairy. The therapist advised Claire to remove these two foods groups completely from her diet. Calcium and vitamin D supplementation was not advised at that time. Claire followed this dietary advice and felt better in the short term; however her symptoms never fully abated.

Life became busy for Claire, and she had three healthy children in quick succession. She breast-fed all her children for a couple of weeks. During these pregnancies she remained on a wheat- and dairy-free diet.

Elaine carried out a full review of Claire's gut symptoms. She had filled out a weekly food and fluid diary. Claire's diet was very high in fibre and fructose. She also had a high intake of fermentable vegetables. Elaine prescribed the FLAT Gut Diet and advised Claire to reintroduce into her diet dairy products such as cheese that have a low lactose content. She continued her food diary for a month.

Claire returned for a review consultation and said 'she felt great' and 'she had a flat tummy, very little bloating and her energy had increased after seven days on the diet'. She loved having cheese back in her diet with no side effects. At this review they discussed gradually reintroducing wheat into her diet and increasing her dairy intake further. Elaine also recommended a low dose of magnesium and vitamin K to help optimize bone health.

Two months later Claire returned for a review. Her IBS symptoms had improved by 80 per cent. The latter 20 per cent she felt was related to her busy and sometimes stressful life, and when she over-consumed fruit and bread. She was very happy with her progress.

Take-home message Dietary restriction based on non-validated food intolerance (sensitivity) test results can have serious health consequences such as osteoporosis. If whole food groups are removed from the diet without proper professional dietetic advice, unnecessary overly restrictive diets can then lead to severe vitamin and mineral deficiencies. Other dietary components may be the cause of your digestive symptoms.

Food intolerances

While food allergies are relatively uncommon, food intolerances are very common and are thought to affect 20 per cent of people.[62] Food allergy is an immune system response, whereas food intolerance is a digestive system response. The definition of food intolerance is a little vague and states that it is 'a non-immunological response caused by a food or food component, at a dose (quantity) that is normally tolerated and which accounts for most of symptoms experienced'. Or, to put it a bit more simply, a food intolerance means that when you eat a certain food, or a certain amount of that food, you experience some unpleasant symptoms. With food intolerance, it is the amount of a food that determines the severity of the symptoms.

Food intolerances are sometimes known by different names, which can be a little confusing. Other names include food sensitivity, or food malabsorption (which means the foods are not absorbed properly) or even food maldigestion (which means that the food in question is not digested properly). But, regardless of the names used, all of them mean the same thing – you don't tolerate a certain food or the quantities eaten are too high (for you) and that leads to unpleasant symptoms.

Symptoms usually occur within 30 minutes to a few hours of eating the food in question, and the common gut symptoms of food intolerances include:

- bloating;
- cramps;
- diarrhoea;
- excessive wind/gas.

As you can see, these symptoms are very similar to those of IBS, which is why diet is very much part and parcel of the management of IBS, as many people with IBS have food intolerances that respond to a reduction in the amount of these foods eaten. Food intolerances lead to unpleasant symptoms, but unlike food allergies, they are never life threatening.

There is no universally recognized way of classifying food intolerances, so we are going to discuss them under two main headings, as follows:

1. Malabsorption of carbohydrates: fermentable carbohydrates (including the special case of wheat intolerance)
2. Sensitivity to naturally occurring food chemicals

1 Malabsorption of carbohydrates: fermentable carbohydrates

A number of carbohydrates or sugars in our diet are not fully digested or absorbed from the small bowel, and are then digested and fermented by the GM in the large bowel leading to gas production, bloating, wind and diarrhoea. These are called fermentable (by the GM) carbohydrates. Although most people do not fully digest these sugars, only those with IBS tend to develop symptoms. A small percentage of people have a genetic lack of a digestive enzyme to fully digest certain sugars such as lactose or sucrose.

Researchers in Monash University, Australia, began to study the combined effect of the quantities of these sugars in the diet on IBS symptoms.[66] They called these fermentable carbohydrates and sugars 'FODMAPs', which stands for **fermentable oligo- di- mono-saccharides and polyols** (that is fructans, galacto-oligosaccharides, lactose, fructose and sugar polyols). FODMAPs have been found to cause digestive symptoms, particularly in people who have IBS.[36,67,68] The low FODMAP diet has become a widely used and effective treatment for many people with IBS, on the basis that some people with IBS have an intolerance to certain carbohydrates. This is an evidence-based, effective dietary strategy to manage IBS symptoms with what has been referred to as a 'top down' approach (i.e. you start at the top, cut out lots of things and then gradually add them in again). In our clinical practice we have incorporated many aspects of this dietary approach but we have always embraced a less restrictive approach, one we refer to as a 'bottom-up' approach. This is the basis of the FLAT Gut Diet and we have found excellent results for our patients with this approach, which is less restrictive and more sustainable.

How carbohydrates cause digestive symptoms

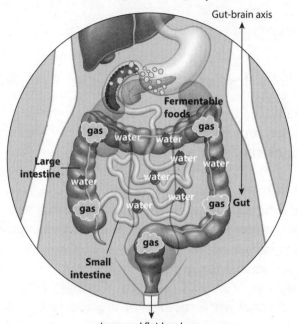

Figure 14 How fermentable carbohydrates cause digestive symptoms

How carbohydrate sugar malabsorption/intolerance causes digestive symptoms	
Fermentation	Sugars not digested in the small intestine enter the colon where they are digested and fermented by the GM leading to production of gases. The gases distend the colon causing bloating, pain and flatulence. The extent of symptoms depends on how much of the particular food you have eaten in one sitting. Think of the composting bin in your garden.
Fluid	The poorly digested sugars draw fluid into the bowel (they are osmotically active), causing distension of the bowel, pain, cramps and diarrhoea.

Fermentable carbohydrates

Now we'll introduce you to the most common fermentable carbohydrates/sugars that are associated with malabsorption or intolerance, and which tend to cause most problems for people with IBS. These include fructans, lactose, fructose, polyols and sucrose. We'll also talk a little more about wheat (which contains fructans), as it is a 'special case' – so many people are deciding to follow a completely wheat- or gluten-free diet, and we want to show you why this simply is not necessary for most people.

i. Fructans (wheat, onions, garlic and leeks)

Fructans are long chains of fructose with a glucose molecule at one end. They are poorly absorbed, and, if present in large amounts in the diet, they are fermented by the gut bacteria. People with IBS are particularly sensitive to the amount of fructans in their diet and have been shown to respond to a low FODMAP diet.[69]

Wheat and other grains are the main sources of fructans in the diet, which is why we decided it deserved a separate discussion. The other main source of fructans in everyday diet is the 'Allium Family', which includes onions, garlic and leeks. Alliums contain much higher concentrations of fructans than wheat, which is why they can have such a potent effect on the gut.

Why do these foods cause problems?
Humans are really poor at breaking down the chains of fructose in fructans, and they are pretty much non-digestible. More than 95 per cent of the fructans we eat are not absorbed from the small intestine, so they reach the colon, where the GM have a feast, leading to all the effects we show in Figure 14. In people with IBS who have the visceral hypersensitivity that we spoke about previously, these effects are amplified: add in the effect of a heightened gut–brain axis and you can easily see why these particular foods cause problems.

The total fructan load at one meal or over the course of the day can cause pronounced digestive symptoms in women with IBS. A wheat pasta dish with a milk-based sauce containing garlic and onions, for example, is a recipe for IBS hell. We will be showing you how to reduce your intake of these foods, in a way that is tasty and sustainable.

Do fructans have benefits?

Yes they do! Remember that they are a form of fibre, and so have many health benefits. It is often a question of simply working out how much of a food you can tolerate, so that you get the benefits without the downsides.

Potential benefits of fructans	
Prebiotic effect	Stimulate the GM
Increase stool bulk	Prevent constipation
Maintain gut barrier	Prevent 'leaky gut'
Lower sugar absorption	Help prevent Type 2 diabetes and reduce obesity
Help calcium absorption	Maintain healthy bones
Lower fat absorption from diet	Prevent high cholesterol
Combination of effects	Reduce risk of colon cancer

Is there a win–win solution?

The good news is that yes there is. The old saying 'everything in moderation' is the answer. We want to get the health benefits of fructans, while reducing the gut side effects. Rather than eliminating these foods from the diet, we want to help you find your individual tolerance level, and we will be focusing on this in Section 5.

Grace's story

Grace, a 38-year-old mother of two, had been diagnosed with IBS and was referred to Elaine for dietary advice. Two years previously she had developed digestive symptoms of bloating, wind and intermittent loose bowels.

Grace was fit and well other than her current digestive symptoms. Both she and her husband loved cooking and had done a wonderful Italian cookery course together two years previously. Grace had tried a gluten-free diet for three months and had not found any improvement in her symptoms. Elaine carried out a detailed review of Grace's diet; it was healthy and varied, and not excessive in lactose or fructose. She had a high intake of vegetables. Discussing her cooking, Grace said that she and her husband used onion and garlic in everything. They loved the flavours and, since doing the cookery course, they were using more than ever, even marinating food with garlic. Elaine explained that onion and garlic have the same components as wheat (fructans) but in stronger amounts. These can produce all the gut problems of which she was complaining. Elaine asked her to remove these completely from her diet for four weeks.

At her review appointment Grace was astounded that her symptoms were gone in a week. She felt much improved and could not believe it was related to onion and garlic. She now used red and green peppers at the start of every meal, where before she would have started with onion and garlic. Grace said her husband was missing the flavour from onion and garlic but she was reluctant to reintroduce them as she felt so well. Elaine advised Grace that she could try to reintroduce them in small amounts in line with the Step 2 – Challenge phase of the FLAT Gut Diet (of which more later) to find her tolerance level. Elaine also suggested that Grace could try some garlic-infused oil in her cooking, as this gives a hint of garlic flavour and is very well tolerated as the actual amounts of garlic in the oil are extremely low. However, Elaine explained that Grace would likely only tolerate these alliums in small amounts.

Take-home message Alliums can cause significant gut symptoms. Garlic-infused oil can add a hint of garlic to cooking without causing gut symptoms.

Fructans and the special case of wheat intolerance

Wheat is often the first thing that people eliminate from their diet when they suspect that they may have a food intolerance and this is particularly the case for people with IBS. *And*, they often feel better as a result. Why is this? Are they gluten intolerant? Usually not.

As we've just explained, wheat is also a rich source of fructans and bran fibre, both of which are fermented by the GM. For this reason many people find that cutting out wheat can reduce bloating. In this instance it is *how much* wheat, or the total 'wheat load' at any one time that is important, and simply reducing wheat intake can be sufficient.

Non-coeliac wheat sensitivity (NCWS)

Many people are now self-reporting and self-diagnosing themselves with gluten sensitivity even when tests for coeliac disease are negative. This was previously called non-coeliac gluten sensitivity, but the name has been changed to NCWS in recognition of the fact that it is usually not gluten, but other things like fructans, that are causing symptoms.[70]

People differ in their tolerance levels to wheat products, before they develop gut symptoms such as bloating. Many people are eating more wheat than ever before and this has the potential to cause problems for some. Individual tolerance level needs to be considered, rather than elimination of a total food group. If someone follows a gluten- or wheat-free diet for a period of time and then consumes wheat, they can experience bloating. This is not evidence of gluten intolerance but is a normal physiological response to the reintroduction of gluten/wheat after a period of restriction, possibly related to increased fermentation of wheat fructans and fibre by the GM.

What's the problem?

We all have freedom of choice, and if someone feels better following a wheat-/gluten-free diet, why not just encourage them to do so? We believe, based on our clinical experience, that many people with IBS tolerate wheat, in measured amounts. You may feel bloated after a large bowl of pasta, but are fine after a medium-sized slice of wholemeal bread. For most people, the benefits of including wheat in their diet outweigh the downsides. But remember, if you have coeliac disease, it is vital to exclude all wheat- and gluten-containing foods.

Four reasons to keep wheat in your diet (*if you are not coeliac*)

- Wheat and wheat-containing products add variety to your diet, and cutting them out will reduce your dietary repertoire.
- Wholegrain wheat foods are an excellent source of fibre, which is good for your GM and good for triggering your gut motility and helping to maintain a regular bowel habit.
- A gluten-free diet has been shown to reduce levels of healthy *Bifidobacteria* in the gut.
- Many gluten-free products are low in fibre and high in sugar and are not actually healthier than wheat-containing products.

Now, let's get back to discussing the other important fermentable carbohydrates...

ii. Lactose intolerance

Lactose is a disaccharide (a double sugar) carbohydrate made up of two linked sugars – glucose and galactose. Lactose occurs naturally in all animal milks, including milk from cows, sheep and goats (and humans). For lactose to be absorbed from the small intestine the link between the two sugars needs to be broken and the single sugars released. The link is cleaved by an enzyme called lactase, which is found in the villi of the small intestine.

Many adults have relatively low levels of lactase enzyme activity, which means that they will incompletely absorb lactose. Lactase deficiency is very common among people of Asian, African and African-American origin, where it can affect 70–90 per cent of people, but affects only a small percentage of Caucasians.[71] Undigested lactose passes to the colon where it is digested by the GM, leading to gas production, bloating, cramps and diarrhoea. The undigested lactose also draws water into the bowel (an osmotic effect) causing further bloating and diarrhoea.

People with lactose intolerance have some lactase activity, and the tolerance level varies from person to person and depends on lactose load. Most people with lactose malabsorption/intolerance can actually tolerate small amounts of lactose (12–15g/day). And on the other side of the coin – most 'normal' individuals have lactose intolerant symptoms when they consume more than 50g of lactose per day. We will discuss the lactose content of many dairy foods in Chapter 20.

In clinical practice we encounter many variations of lactose tolerance. Most patients who develop symptoms tend to have a very high lactose load in one sitting – for example cereal with milk, which is then topped off with yogurt. This can exceed their lactose tolerance load, producing symptoms approximately one to two hours after eating.

Secondary lactose intolerance

Sometimes people can develop 'secondary lactose intolerance' when they have another condition that damages the villi in the small bowel, where the lactase is located. The most common examples are coeliac disease, Crohn's disease or an episode of severe gastroenteritis. In all situations the lactase enzyme activity recovers once the underlying condition is treated/settles down. As we get older, we tend to lose more lactase activity in the lining of our gut, so that older people are also more prone to lactose intolerance.

Hidden sources of lactose

Lactose is often used as a bulking agent in pills. If people (particularly the elderly) are taking a lot of medications, the lactose in the medication can cause symptoms of lactose intolerance.

Treatment of lactose intolerance/malabsorption

It is important to have a proper diagnosis. People who are lactose intolerant, can usually tolerate lactose in small amounts, particularly if spread out over the day (12–15g divided into three portions) and a low lactose diet, rather than a lactose-free diet is usually sufficient. They can consume lactose-free foods and foods low in lactose such as cheeses, butter and cream. They can tolerate small amounts of milk, yoghurt, ice cream and chocolate. It is important to note that most cheeses are very low in lactose (high in protein and fat but low in lactose) and so can be safely included in a low lactose diet. This is particularly important for bone health, as cheese is an excellent source of calcium and vitamin D.

Lactase enzyme supplements can be taken in tablet form to aid lactose digestion and reduce gastrointestinal side effects associated with consumption. So, if you really want an ice cream on a warm summer's day, lactase supplements will help you to enjoy your treat!

Diagnosis of lactose intolerance

1. The diagnosis is generally reached from taking a detailed dietary history. People have often identified a link themselves between milk products and their symptoms.
2. A lactose breath test can also be carried out.

Dairy-free diets and your bone health

If you follow a dairy-free diet, you are at increased risk of calcium and vitamin D deficiency due to potential restriction of calcium-dense foods. It is vital to ensure the calcium content of your diet is adequate, either through the food you eat, or through use of supplements if necessary.

iii. Fructose intolerance (malabsorption)

Fructose is a single sugar (monosaccharide). It is the natural sugar in fruits, honey and some vegetables.

Despite all the health benefits of fruit, fructose intolerance is quite common and affects between one-third and two-thirds of the population according to different studies.[72] So, it is actually possible to have too much of a good thing. In the USA in particular, there has been a big increase in the use of fructose in the form of high-fructose corn syrup, as a sweetener in many foods and beverages.

Fructose malabsorption occurs when our small intestine fails to absorb all the fructose in the diet and the excess spills over into the colon where it acts as 'dessert' for the GM. This produces gas, bloating, cramps, flatulence and diarrhoea, as we explained above.

People vary hugely in their ability to absorb fructose, and symptoms of fructose malabsorption depend on the 'fructose load' eaten at any one time. Fructose malabsorption is a significant trigger for IBS symptoms, and we will be looking at this in detail in Step 1 (Simplify phase) of the FLAT Gut Diet.

Free fructose: The balance of fructose and glucose in a food is important

When fructose is present in the intestine in similar quantities to glucose, it is well absorbed, but when fructose is in excess of glucose it is poorly absorbed. The excess of fructose over glucose is called 'free fructose' and this is the part that is generally poorly absorbed and responsible for causing unpleasant symptoms in some people. Below are some interesting examples of the fructose/glucose balance of some different fruits.

Fruit	Fructose g/100g fruit	Glucose g/100g fruit	Free fructose	Resulting absorption
Apple	7	2	5	Poor
Pear	7	3	4	Poor
Orange	2.0	2	None	Good
Rhubarb	0.4	0.4	None	Good
Honey	41.8	34.6	7.2	Poor

Diagnosis of fructose malabsorption
A fructose breath test is used to detect fructose malabsorption.

The fruit smoothie combo!
If you eat fructose along with lactose, the fructose/lactose load eaten at one sitting can cause bloating, excessive wind/gas and loose stools. The

breakfast fruit smoothie (perhaps with added honey, which has a high free fructose level), which has become part of a healthy routine for many people, can unwittingly cause digestive symptoms for some people.

Elaine's fruit smoothie incident!

When smoothies first came onto the market, Elaine purchased a small raspberry smoothie, and when she got home late from work she left it on a table beside the couch and then forgot to drink it. The next morning, when she walked into the sitting room she was greeted by a vision 'like a ketchup murder scene on TV'! There was raspberry pulp everywhere – ceilings, fireplace, walls and carpet. The far-reaching explosion from 150ml of juice was unbelievable. She contacted the company to ask if this was usual as it was a new product to the market. They said they were experiencing problems with juice exploding, due to a fault in the packaging, which did not have a breathable lid to release the gas produced from fruit sugar. Similar side effects may be observed when you drink high fibre smoothies with fermentable fruit and vegetables – you can draw your own conclusions about what is happening internally.

Alice's story

Alice, 24, was referred to Elaine's clinic. She was a lively and engaging young woman, with a healthy weight of 61kg (134lb) and a BMI of 22. She was a non-smoker and drank two units of white wine weekly. She was active and walked to and from work and exercised in the gym three times per week. Alice had no significant personal medical history but her brother had been diagnosed with Crohn's disease at the age of 17. Alice had been thoroughly investigated by Barbara, who had diagnosed her with IBS and had referred her to Elaine for dietary management.

Elaine explored her symptoms and daily eating patterns. Alice explained that she generally felt well first thing in the morning, but from lunchtime onwards her abdominal bloating would start to build. She had to loosen her clothes in the evening after dinner and showed Elaine a photo on her phone of her distended tummy 'almost like I am pregnant'. She passed two or three loose bowel motions per day, sometimes with pungent odour, and these tended to start in the afternoon going on into the evening. Sometimes she struggled to reach the bathroom on time. She was tearful and said it was embarrassing and was affecting her quality of life. She and her boyfriend had lived together for the past two years and were very healthy eaters. He was very health conscious and had a personal trainer.

Alice had recorded her food and fluid intake for a week prior to attending the appointment. She was proud of her healthy diet. When asked if she had altered her diet in recent years, she said that she and her boyfriend were eating healthier since moving in together two years ago. Their treat foods were homemade date and fig protein balls in the evening and popcorn several times during the week. Together she and Elaine went through her diet.

For breakfast, Alice had a homemade fruit smoothie with porridge oats, mixed berries, a kiwi and about 200ml orange juice. Mid-morning snack was an apple and an orange. Afternoon snack was about ten grapes. Then in the evening she had her date and fig protein balls and would pick at some more grapes and sometimes ate popcorn.

Alice's daily fructose intake was 29g per day.

Elaine explained to Alice that she was likely exhibiting symptoms of fructose malabsorption. Her boyfriend who was also consuming large amounts of fructose obviously had a higher threshold, as he had no symptoms. Not only was Alice's daily fructose very high, it was also too high at one sitting. She recommended that Alice reduce her fructose load to 12 grams per day (3 x 4 grams per meal) and no more than 4 grams per meal.

Alice attended for a review three weeks later and described the diet as 'life changing'. Her symptoms had resolved within a week. She was so relieved as she really thought it was something quite serious causing her gut problems. It never occurred to Alice that she could be eating 'too healthily'. Elaine ensured that Alice's diet was nutritionally balanced and advised her regarding a gradual increase of her fructose load to find her individual threshold for fructose tolerance.

Take-home message Healthy foods taken in large quantities can cause significant gut symptoms. A high fructose load may not be tolerated by women (or men) with IBS.

iv. Polyols

Polyols are sugar alcohols and occur naturally in foods. Sugar polyols (the giveaway is that they almost all end in 'ol') including sorbitol, mannitol, xylitol, maltitol and isomalt (the odd one out) are poorly absorbed. These substances are widely used nowadays as sugar substitutes in many 'diet' and 'low-fat' products (including sugar-free chewing gum). They also occur naturally in some vegetables including mushrooms and avocado. When we eat foods containing polyols, only approximately 20 per cent of the polyols are absorbed by the small intestine and the rest enters the colon where they have an osmotic effect – this means that they draw water into the colon – and as a result, can cause diarrhoea and bloating.

v. Sucrose malabsorption

Sucrose (table sugar) is a double sugar (disaccharide), made up of a glucose molecule linked to a fructose molecule. Sucrose is normally broken down in the small intestine by the enzyme sucrose-isomaltase and most of us have plenty of this enzyme. Previously it was thought that sucrase-isomaltase deficiency (SID) was rare, but more recent studies suggest that 2–9 per cent of Caucasians in the USA have genetic variants that may cause some degree of sucrose malabsorption.[73] The frequencies of genetic variants among people of Asian and African-American descent are not known. The symptoms are very similar to those of lactose and fructose intolerance and can mimic IBS, causing diarrhoea, bloating and excessive gas. It can be diagnosed using a sucrose breath test.

2 Sensitivity to naturally occurring food chemicals

Foods contain many other contents and chemicals, which can cause reactions in some people. The most common of these are briefly mentioned below.

Sulphite allergy/sensitivity

Sulphites are naturally occurring chemicals that are found in fermented foods and drinks such as wine, cider and beers. They are also found in dried fruits and ready prepared sauces, as food additives and in medications. Sulphite allergy/sensitivity is uncommon and tends to occur mainly in people with asthma and other allergies. Studies suggest that up to five per cent of asthmatics may have this sensitivity, compared to one per cent of the 'normal' population. Most reactions to sulphites are characterized by spasm of the airways (choking), which can occur within minutes of eating sulphite-containing foods. Contrary to what we said about food intolerances not being life-threatening, this is one example of a non-allergic food intolerance that can cause severe effects.

Histamine intolerance

Histamine is a chemical that is naturally produced in our body and is also present in high levels in certain foods. Some people have low levels of the enzymes that deactivate histamine and the food-derived histamine can activate mast cells, a type of immune cell in the gut. Activation of the mast cells leads to the release of other inflammatory proteins into the bloodstream that can mimic an allergic reaction and can cause migraines, headaches, diarrhoea, flushing, hives, eczema and rhinitis (hay fever). This is called histamine intolerance, and it is poorly understood and often very difficult to diagnose.

High levels of histamine are found in alcohol (particularly red wine), pickled foods like sauerkraut, vinegar, mature cheeses, shellfish, smoked meat products, fermented soy products, nuts, beans, pulses, chocolates, ready meals, and salty snacks. Foods can also trigger release of histamine within the body, and these include most citrus fruits, plums, straw-berries, bananas, alcohol, avocados, chocolate, shellfish, nuts, cow's milk, tomatoes, wheatgerm and many artificial preservatives and dyes.

Treatment of histamine intolerance involves cutting down on histamine intake and triggers. As you can see, these are very long lists of foods, and as a result, a low histamine diet is very restrictive. Some enzyme supplements are also available commercially to help break down histamine; one of these is called diamine oxidase (DAO).

Other food chemicals

There are many other chemicals present in foods, either naturally occurring, or present as additives and preservatives, which, in rare cases, can cause digestive symptoms or intolerance in some people. A discussion of all of these chemicals is beyond the scope of this book, but if you are having significant symptoms, we would urge you to discuss this with your GP/family doctor and perhaps seek referral to an allergist or an immunologist.

So what does all this tell *you*?

Food allergies are relatively uncommon, but are more common in people who have other allergic conditions such as asthma, eczema or hay fever. If you have other allergic conditions and feel that foods may be triggering digestive symptoms, it would certainly be worthwhile seeing a medical immunologist or allergist, as diagnosis of food allergies can be complicated, particularly the delayed-type of reaction we described earlier. If you do not have allergies, then a specific food allergy is unlikely, although not impossible, and food intolerance is much more likely. Commercial and alternative practice 'food allergy testing' is misleading and can lead to overly and unnecessarily restricted eating, so we strongly advise against their use.

Food intolerance is now known to play a large role in triggering symptoms in people who have IBS, and also functional dyspepsia. With any food intolerance, it is *how much* of a food you eat at a given time that is important, and you do not have to cut out these foods completely. This is important when it comes to sticking to a dietary approach in the long term. If you know that you can have your favourite pasta dish, or an ice-cream or a serving of crisp, cold, fresh watermelon from time to time, you will find things a lot easier than you would if you were told never to eat any of your favourite foods again.

The FLAT Gut Diet does just this: it finds your individual threshold for many different foods, allowing you to eat a varied, interesting and healthy diet that you can enjoy for the long haul.

Take-home messages

- True food allergies are uncommon in adults.
- Food intolerances are common, particularly in people with IBS.
- Most common food intolerances are caused by fermentable carbohydrates.
- With food intolerance is it *how much* of the food you eat that is important, and you need to find your own individual tolerance level.
- Most people do not need to cut out dairy (lactose) from their diet.
- Contrary to what people believe, most cheeses contain almost no lactose.
- Most people can tolerate wheat, you just need to find your tolerance level.
- Commercial 'food intolerance' tests are of no value and lead to unnecessarily restricted diets. Don't waste your money on these expensive tests.

13

Alcohol and your gut

Many people enjoy alcohol in moderation. However there is no doubt that – increasingly – some people drink more than their body likes or tolerates. We listen to our bodies in so many ways, and yet sometimes people do not listen to their bodies when it comes to alcohol intake. We are more affluent than previous generations and have access to a wider and more enticing range of alcoholic drinks, and people, without realizing it, may be drinking more than is good for them. Recent studies show that women are drinking more than ever before.[74]

What is a unit of alcohol?

What exactly is a unit of alcohol? Technically speaking one unit of alcohol equals 10ml or 8g of pure alcohol. Countries can differ in their terminology, and some talk about 'units', whereas others talk about 'standard drinks'. The amount of alcohol in these measures varies between 8–14g. What this translates into in real terms can often shock people – that glass of wine you drink at home in the evenings may contain 2–3 units of alcohol, not just one. The number of units you are drinking can sometimes be double or even treble what you thought.

How does alcohol affect your digestive system?

Alcohol can affect every part of your digestive system:

The effects of alcohol on the digestive system	
Organ	Effect
Liver	Inflammation (alcohol-related hepatitis) Fatty liver Cirrhosis
Pancreas	Acute pancreatitis Chronic pancreatitis
Oesophagus	Increased acid reflux and GERD Increased risk of oesophageal cancer
Stomach	Gastritis Worsening of functional dyspepsia
Small bowel	Increased motility and gut permeability (leakiness) Worsening of IBS symptoms Diarrhoea
Large bowel	Worsening of IBS symptoms Diarrhoea

The liver

The female liver weighs about 1.4kg, about 30 per cent smaller than the male liver, which weighs on average 1.8kg. This means that we have 30 per cent fewer detoxifying enzymes, including those enzymes involved in metabolizing alcohol. What's more, women are affected differently by alcohol, and not just because of the size of our livers. Women produce less of the enzyme called alcohol dehydrogenase, which means that we break down alcohol less efficiently than men. Alcohol tends to be retained in fat, and water tends to disperse it, so because women tend to have more body fat than men, we are more sensitive to its effects.

When we exceed our capacity to metabolize alcohol, toxic inter-mediates are produced in the liver, and they damage the liver cells (known as hepatocytes). In the short term this leads to irritation and inflammation of the hepatocytes (this is called alcohol-related hepatitis). This causes the liver enzymes, which are measured in the blood, to rise. Over the medium term, alcohol-related irritation of the liver can lead to 'fatty' liver, where the liver cells become filled with fat. Over a longer period of time, the alcoholic hepatitis and fatty liver can lead to extreme scarring and fibrosis of the liver. This end stage of damage is known as cirrhosis. Figure 15 shows us the difference between a healthy and a diseased liver.

The liver is a remarkable organ, designed to metabolize and detoxify all the substances we eat and drink. Up to 90 per cent of the liver can be damaged before people begin to show signs of cirrhosis or liver failure, and at this stage unfortunately it is usually too late to reverse the damage. The signs of liver failure include a swollen abdomen and yellow discolouration of the skin and eyes, which is called jaundice. While having so much spare capacity is a wonderful in-built safety system, it also means that some people only realize the extent of their liver damage very late. At this point, all you can hope for is to prevent the remaining liver function from deteriorating further.

Any gastroenterologist in the western world will tell you that over the last 15 years we have seen an increase in the number of women in their thirties presenting with signs of liver failure secondary to alcohol excess. These are not necessarily women you would think of as 'alcoholic'. Many of them are working and raising a family but they have been drinking too much and too frequently. Men who have been heavy drinkers tend to develop cirrhosis much later, in their forties or fifties.

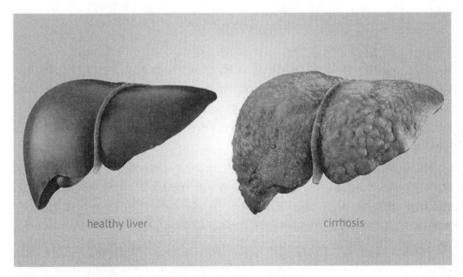

healthy liver　　　　　　　　　　　　cirrhosis

Figure 15 A healthy liver and a liver affected by cirrhosis

The pancreas

The pancreas is exquisitely sensitive to excess alcohol. An acute excess of alcohol can cause the pancreas to become severely inflamed: this is called acute pancreatitis. This is a very painful and potentially life-threatening condition and usually requires hospitalization. Chronic pancreatitis is a result of damage to the pancreas over a longer period. Alcohol is one of the more common causes, but not the only cause, as people can have a genetic predisposition also. Chronic pancreatitis causes pain and a lack of the pancreatic digestive enzymes, leading to diarrhoea and nutritional deficiencies. It can also cause diabetes.

The oesophagus

Alcohol can be a direct chemical irritant to the oesophagus, causing inflammation and discomfort. It relaxes the lower oesophageal sphincter, allowing acid and stomach content to reflux up from the stomach into the oesophagus. This is why alcohol can cause heartburn.

The stomach

Alcohol can act as a chemical irritant to the stomach, causing gastritis and ulcers. Even in the absence of visible irritation, alcohol can trigger symptoms of functional dyspepsia in some people.

The large and small intestines and IBS

We will mention these together as the effects are similar. Alcohol can stimulate gut motility, leading to diarrhoea. It has also been shown to increase intestinal permeability (leaky gut) and all the associated problems. Chronic alcohol use is associated with an alteration in gut bacteria in the colon, as well as affecting reabsorption of water from the colon, also causing diarrhoea.

If you have IBS, these effects all tend to be exaggerated, so that cramps, bloating and diarrhoea may be worse than those of people who do not.

Is one type of alcohol worse than another?

In terms of the potential for alcohol to damage the liver or the pancreas, it does not matter which type of alcohol you are drinking, it is the total number of units that is the critical factor. Our patients often say to us 'I was only drinking wine, or beer, no spirits', but that doesn't matter if you're drinking more than is healthy.

Alcoholic drinks, of course, contain other substances that have the potential to aggravate the gut in different ways and certain drinks can have more marked digestive symptoms that others. For example, rum and cider both have a high fructose content and should probably be avoided if you have IBS. Beer contains wheat and fructans, as well as carbonated bubbles and can cause considerable bloating, flatulence and pain. Red wine contains substances called bio-active amines that can cause diarrhoea. And don't forget that all alcohol contains calories and will contribute to weight gain.

Substances in alcoholic drinks that can contribute to gut symptoms	
Additional substances found in alcoholic drinks	Symptoms they can cause
Fructose	Bloating Diarrhoea Cramps
Glucose	Weight gain
Carbonated bubbles	Bloating Flatulence
Wheat	Bloating Flatulence
Yeast	Bloating
Calories	Weight gain
Sulphites	Nausea
Bioactive amines (histamine and tyramine)	Altered motility, possible increased permeability Diarrhoea, bloating, cramps

How much alcohol is safe to drink?

The World Health Organisation (WHO) does not set particular limits for 'safe' alcohol consumption. Indeed in recent times, the term 'safe' drinking has been replaced by 'low-risk' alcohol intake, as the WHO believes that the best solution for our health is not to drink at all.

Essentially, the less one drinks the better. It's important to realise that all 'low risk' guidelines tell us what the upper limit is, and this doesn't mean that we should drink that much, and certainly not on a regular basis. Remember also that 'home poured' alcohol measures tend to be bigger than you realise.

It can be very difficult to interpret how much alcohol is deemed 'safe' to drink, as different countries measure and talk about alcohol in seemingly different ways. But actually most countries have very similar recommendations for low-risk or 'safe' drinking.

For example, in Ireland, alcohol is discussed in terms of 'standard drinks', and a standard drink in Ireland contains 10g of alcohol. This equates to a small glass of wine (100mls). The safe drinking limit for women is 11 drinks per week, which is a total of 110g of alcohol per week.

In the US, they also talk about 'standard drinks' or measures, but the size is bigger than that in Ireland, and a standard drink is 14g of alcohol, which equates to just under 150mls (5 fluid ounces) of wine or one shot of spirits. The weekly safe drinking guideline in the US is eight standard drinks for women, which is 112g of alcohol per week.

In the UK, alcohol is discussed in terms of 'units' and a unit of alcohol is measured at 8g. Women are advised not to drink more than 14 units per week, which is 112g per week.

Aside from the liver and digestive effects of alcohol, regular alcohol consumption, regardless of how much, is associated with an increased risk of certain cancers (even just one drink per day has been shown to marginally increase the risk of breast cancer) and carries an increased risk of long-term damage to the neurones in our brain and more chance of developing dementia or cognitive decline as we get older.

From the gut health point of view – *listen to your body*. If you get diarrhoea, bloating, nausea or abdominal cramps after drinking alcohol, then you are drinking more than *your* body likes.

And remember – in the background, your liver bears the brunt of all the alcohol you drink and ultimately will come under pressure.

Ava's story

Ava, 23 years old when she was referred to Barbara, was generally very healthy and was studying business at university and living in student accommodation. In general, she had mild IBS-type symptoms, but managed these through diet and was doing well. At university, Thursday nights tended to be the 'big night out', when lots of students would party and drink. She and her friends would usually have 'pre-drinks' and then go out and have some more. She said that she could have ten to 12 drinks on a Thursday night. She went home about once per month, and when she did, she tended go out on a Saturday night with some school friends, and would do the same thing.

Ava's problem was that every time she went out drinking with friends, the following day, she had severe nausea and vomiting and between six and eight bouts of diarrhoea. This would last a day (she often missed lectures on a Friday) and then she would feel fine again. Ava wanted to know how to stop the GI symptoms on a Friday. She was concerned that there was something serious underlying them, as her none of her friends had this problem.

We spoke about her drinking pattern and how this was likely aggravating her otherwise mild IBS symptoms. Barbara explained that people differ in their sensitivity to alcohol, and how the 'binge' drinking pattern was clearly a problem for her digestive system because she more than likely had underlying IBS. Ava felt that her drinking pattern was not excessive and that this was a very important and enjoyable part of her social scene. She was hoping that there might be a medication that could prevent this from happening. We spoke about the very enjoyable effects of alcohol and how consuming fewer drinks would alleviate the symptoms and still allow her to socialize with her friends. Some tests were carried out (some bloods and a gastroscopy, just to ensure that she didn't have an ulcer or GERD), all of which were normal. We chatted about things again after Ava's procedure, but she didn't come back for review.

Take-home message There is not a pill for every ill, and sometimes our body is telling us the answer that we may not wish to hear. In this case, clearly Ava's binge drinking was having a negative effect on her digestive system, and ultimately the only cure for this would be to reduce her alcohol intake.

Take-home messages

- You should not drink more than the safe-drinking guidelines for your country per week.
- You should not drink every day. Have two or three alcohol-free days per week.
- Try not to drink more than 6 units in any one sitting.
- If you are having more than 3–4 units, you should try to drink this with food.
- The less you drink, the lower the risk of alcohol-related health problems.
- Listen to your body.
- Remember your liver is in the background working away to detoxify the alcohol you drink, and can be under pressure without you knowing it.
- If you have IBS, you should avoid rum, cider, dessert wine and cocktails. Other alcohols can be tolerated in moderation (see Chapter 20).
- If you have functional dyspepsia, all alcohols can be a trigger, and should be kept to a minimum.

Section 4

SOME EASY WINS: FIRST STEPS TO SOLVING YOUR GUT PROBLEMS

14

Your first steps in managing IBS

A quick recap

Well done! You've read a lot of information to get to this point and by now you should have a much better understanding of your IBS. Just a quick reminder of some of the top-line points:

- **Personalized treatment:** The management of IBS must be personalized, and is not a case of 'one size fits all'. As we talked about in Chapter 5, someone with constipation-predominant IBS (IBS-C) will need a very different approach to management than someone with diarrhoea-predominant IBS (IBS-D). However, even more challenging is the fact that many people with IBS actually fall into the mixed or alternating type of IBS (IBS-M) or the unspecified type.
- **Management, but not a cure (at least not yet):** At this point in time there is no 'cure' for IBS, but there are many, many things than can be done to help your symptoms and help you lead as normal and pain-free a life as possible.
- **Hormones:** We spoke about this in Chapter 7. If your menstrual cycle has a significant impact on your IBS symptoms, have another look at the simple steps we suggested in Chapter 7.
- **Find the right clinician:** It's vital that you find a clinician (doctor or dietitian) that you connect with. It has been scientifically proven that your relationship with your clinician plays a large role in your response to treatment.[26] This makes sense; if you feel that your condition and concerns are not being taken seriously, or that you have not been appropriately investigated, or that your physician is viewing you in a negative light (anxious, overly emotional or unbalanced in any way), it's not a recipe for success. There needs to be mutual trust and openness.
- **Appropriate tests:** We spoke about the importance of getting the right tests and the correct diagnosis before starting out on any

treatment plan. Make sure that you have done this and that other conditions have been ruled out.

- **Holistic approach:** Successful management of IBS involves a holistic, step-wise approach that addresses every aspect of your wellbeing. We spoke earlier about the gut–brain axis, and you may know from personal experience about the significant role played by the mind and emotions in your symptoms. We call this the TEAMS approach (see below).

For many people, making some fairly simple changes to diet and lifestyle can solve their IBS symptoms. For others, unfortunately, the solutions may not be so simple, and if you're reading this book, you may fall into this more complex category. We're going to help you find your solutions, through a 4-step plan. You may just need to make one or two simple changes, or you might need to follow all four steps, or parts of all of them, to find your solution. The next section of the book, Section 5, is all about the FLAT Gut Diet, and we know that you're impatient to get to this, but first, let's make sure that you've got the basics right.

The Gut Experts' 4-Step IBS Solution

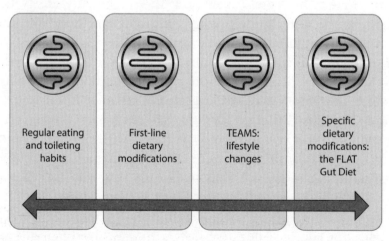

| Regular eating and toileting habits | First-line dietary modifications | TEAMS: lifestyle changes | Specific dietary modifications: the FLAT Gut Diet |

Step 1: Regular eating and toileting habits

Our body likes predictability and routine. Adopting a regular eating pattern, taking time when eating and avoiding too much snacking between meals, can significantly help your IBS symptoms. It is not so simple for everyone unfortunately, but a regular eating pattern should still form the basis of whatever plan you follow. Remember, in Chapter 2 we spoke about the MMC, or housekeeping waves of motility, that only occur when fasting – if we never fast, these waves never get a chance to do their thing.

We have found that constipation is much more common among certain professions such as police officers, nurses, doctors, shift workers and schoolteachers. This is often the result of being unable to leave one's work duties to answer the call of nature, and lack of access to bathroom facilities during work so that the natural bowel response is often overruled. When we postpone going to the bathroom in this way, we override the natural sensitivity of the full rectum, and over time this leads to this reflex becoming weakened. Ideally, we should listen and respond to our own body's cues. So whenever possible *listen to the call of nature*.

Step 2: First-line dietary modifications

First-line dietary measures for constipation-predominant IBS (IBS-C)

Fibre: Make sure that you have adequate fibre intake. Generally 20–35g per day is recommended (Chapter 11). Remember, you get fibre from plant-based foods; diversity and 'eating the rainbow' are key. We recommend working out your intake, and if you're not eating enough fibre, to increase by 5g per week.
Some easy sources of fibre:

- 2 kiwi fruits per day can help mild constipation.
- Linseed and flaxseed are good sources of fibre, which is found in the seed coat. Take 6–24g per day – one heaped teaspoon to two table-spoons – of whole or ground linseed or flaxseed. Gradually increase over two–three weeks. They can be added to porridge, cereals, yogurts, soups and salads. They need to be taken with plenty of fluid as they act by absorbing water and adding to the bulk of the stool.
- Psyllium husk, also known as Ispaghula, is a natural soluble gel-forming fibre. It helps relieve constipation. This can be purchased as whole psyllium husks, psyllium husk powder or in capsules. This fibre does not generally cause bloating. It is important to remember that too much soluble fibre, particularly if not taken with enough fluid, can worsen symptoms of constipation. Do remember to drink your fluid.

(Continued)

Avoid high protein diet/low carbohydrate diets as they slow bowel transit causing constipation.

Fluid: Drink 1.5–2L of fluid per day (Chapter 10). Water is the best source, but herbal teas and decaffeinated tea and coffee can contribute to daily fluid intakes. Constipation can be worsened by a very high intake of fibre and inadequate volume of fluids.

Caffeine acts as a bowel stimulant and can speed up bowel transit. Strong tea and strong coffee and 80–90 per cent dark chocolate, can aid constipation.

Sorbitol: Prunes and apricots contain sorbitol, a good natural laxative, but be aware that they can cause bloating and flatulence in some people.

First-line dietary measures for diarrhoea-predominant IBS (IBS-D)

Diarrhoea and cramps can be caused quite simply by eating too much fibre, or fruit or too many chemicals that stimulate the bowel, such as caffeine. So before jumping into very specific dietary forensics, look at your consumption of the following:

Caffeine: Reduce your caffeine intake significantly to see if this improves your symptoms. Remember caffeine is present in tea as well as coffee, and if you are a tea-drinker, you'd be amazed how the caffeine can mount up.

Fibre: Be fibre aware! Is your intake too high for you? If so, try reducing it to around 18 to 20g of fibre per day, then gradually increase levels to find your tolerance level.

Sweeteners: Avoid any sweeteners containing 'ol' at the end in particular xylitol, sorbitol and foods containing them.

Stone fruits: Avoid stone fruits such as apricots, avocado, cherries, peaches, plums and prunes. These all contain high levels of 'ols'.

Avoid hypertonic drinks or drinks with a very high sugar content.

Onions and garlic: Cut these out of your diet to see if this helps. They are a potent source of fructans and can trigger diarrhoea.

Fructose: Reduce your intake of fruit to three small pieces per day.

Lactose: Reduce your lactose load to three portions per day, as recommended in the FLAT Gut Diet (Chapter 20).

Alcohol: should be kept to a minimum.

Chewing gum: Many sugar-free chewing gums contain sorbitol, a FODMAP, which can trigger diarrhoea.

Low-fat products: If you eat these, be aware that some contain sorbitol and other 'ol' sweeteners, which can trigger diarrhoea. Monitor your symptoms after eating low-fat produce.

First-line dietary measures to reduce bloating and flatulence

Because bloating is such a common problem, particularly in women with IBS, we're discussing it separately in Chapter 15, coming next.

Step 3: TEAMS: lifestyle measures

In Chapter 23, we give you specific advice and goals on how to optimize the TEAMS factors.

T	Total mind and body health
E	**Exercise:** Regular exercise can help reduce symptoms of IBS. Aim to exercise 5 times per week for 30 minutes.
A	**Alcohol:** Alcohol can worsen many digestive conditions, including IBS and FD, and intake should be kept to a minimum. If you wish to enjoy alcohol, choose a drink that is less likely to trigger symptoms (Chapter 13).
M	**Mental health:** Mental health needs to be actively nurtured, as stress is often a factor in triggering symptoms.
S	**Sleep:** It's important to get adequate, quality sleep.

Exercise

Exercise and physical fitness are key to maintaining physical and mental health wellbeing. This does not mean that you have to be a regular at the gym, but regular exercise, particularly walking, can be very helpful for people with IBS, and can help stimulate gut motility, particularly in patients with IBS-C. More vigorous exercise and structured exercise programmes have shown significant improvement in IBS symptoms – particularly constipation and overall wellbeing. You should aim to walk for 30 minutes, or do some other form of exercise at least five times per week.

We realize that exercise can be a problem if you have diarrhoea or urgency, as going out to walk, or more vigorous activity can create a lot of anxiety. People often fear 'having an accident' if they need to go to the bathroom, but are not close to bathroom facilities. The key here is to gain better control of the diarrhoeal symptoms first through dietary measures and/or medications and then, once those symptoms are better controlled, to try some gentle exercise. If walking, it can make sense at the start to choose somewhere where you know the location of bathroom facilities, or else close to your own home, just to ease anxiety levels. This is a prime example of where the gut–brain axis can go into overdrive.

Alcohol

In Chapter 13 we spoke about the many and varied effects that alcohol can have on your gut. It is a trigger for many people with IBS and also for those with functional dyspepsia. The effect of alcohol is usually not 'all or nothing' and it is usually possible to enjoy alcohol in moderation.

Mental health

Stress and anxiety are well-recognized IBS triggers, so it makes perfect sense that helping someone deal with stress and anxiety issues might prove helpful in reducing their IBS symptoms. Both cognitive behavioural therapy and hypnotherapy CAN improve IBS symptoms, as can acceptance and commitment therapy (ACT). We're well aware that therapy is expensive and may be hard to find in your area: any method that helps you relax and de-stress is of benefit. Mindfulness or yoga can help with this, and enable you to listen to and be aware of your body. Medications for anxiety and depression may also be necessary in addition to these strategies in some cases, and can be highly effective.[75]

Sleep

Adequate sleep is vitally important for our general wellbeing. Up to 50 per cent of people with IBS report having sleep disturbances.[76] A small study of women with IBS found that self-reported sleep disturbances were associated with increased abdominal pain, anxiety and fatigue the following day. Certainly, IBS symptoms can potentially prevent someone having a restful night's sleep but recent evidence also suggests that sleep disruption may directly enhance visceral hypersensitivity and GI symptoms. It is also true that co-existing anxiety or mood disorder, both of which are common in IBS, can also affect sleep quality. Whatever the mechanisms, restful and so-called restorative (meaning you waking up feeling revived) sleep is important and if poor sleep quality or 'sleep hygiene' is a problem, then this should be addressed.

Step 4: Specific dietary modifications

If the more simple, first-line dietary and lifestyle changes do not bring your symptoms under good enough control, then it is time to move to a more specific dietary plan. An appointment with a registered dietitian can be invaluable. We are introducing the FLAT Gut Diet in the next section, and will bring you through it step by step. You may also wish to try other dietary plans, such as the low FODMAP diet. A lot has been written about this and there are many resources available. The FLAT Gut dietary approach is less restrictive, more inclusive and we are thrilled with the excellent results we have been getting from it for a number of years. Go straight to Chapter 17 to get started.

Medications and probiotics

There are a myriad of medications that can be used in the treatment of IBS. These should never be used alone for the management of IBS, but instead should always be used in combination with the lifestyle and dietary measures that we have just discussed. Many medications are now available without prescription, so-called over-the-counter medications, such as peppermint oil and simple laxatives. Others, such as certain laxatives or anti-spasmodic medications will commonly be prescribed by your family doctor/GP. There are many other medications that are generally prescribed by a specialist, such as a gastroenterologist. The use of medication for treatment of IBS should obviously be under the supervision of a doctor. Medications are chosen to help with certain symptoms and this should be done on a person-by-person basis.

If you are interested in getting more information on the different medications that are used to treat IBS, we have included some of the more common ones in Appendix 1.

We talk about the role of probiotics in IBS in Chapter 27. There is little convincing evidence to support the use of probiotics in IBS and only one-in-seven people will get some benefit. If you are keen to try one, take a multi-strain probiotic for a month, but stop taking it if you do not notice any improvement. Combination probiotics rather than single strains seem most effective, and in particular may help with bloating and flatulence.

Conclusion

Hopefully it is now clear to you that the treatment of IBS must be personalized and there is absolutely no one-size that fits all! By careful discussion with your doctor and dietitian, the different symptoms can be teased out and the best approach for *you* and a management plan for *you* can be put in place. While your IBS cannot be cured, it can be managed, and with the right approach, you can enjoy a significant improvement in your symptoms and quality of life.

15

Why are you so bloated and how do you beat it?

Nearly all of us have experienced bloating, or a gassy, distended abdomen from time to time. But thankfully for most of us, this is just a temporary thing that goes away. For other people, particularly those with IBS, functional dyspepsia and other DGBIs, bloating is a recurrent and severe problem that has a significant negative impact on their daily lives and stops them from doing normal daily activities. If you experience bloating, then you know just how uncomfortable and frustrating this symptom can be.

What is bloating?

We have spoken about bloating briefly in earlier chapters, but because this is such a big problem for many women in particular, we thought that it deserved a chapter all to itself!

As previously mentioned, bloating is a sensation described by many patients (particularly those with IBS) and means a sense of fullness in the abdomen. It is often associated with visible distension of the abdomen, which can feel very hard and tense. Bloating is what someone *feels*, and distension is how the abdomen *looks*. Many women will say 'I look like I'm six months pregnant!' The sense of bloating can be eased by passing wind or a bowel motion, but usually not completely. The bloating often goes down over night and is least marked in the morning, only to build up over the course of the day. Up to 90 per cent of people with IBS experience bloating, but it also occurs in up to 30 per cent of people who do not have a diagnosis of IBS. So this is a very common problem.

What causes bloating?

This is a tricky one as there is no one single cause, and usually a combination of factors contribute to the feeling of bloating. Most people believe that their symptoms are simply due to a build up of gas in their gut, but actually research shows that excessive gas is the cause in only a minority of cases.[77] Bloating has already been mentioned as a feature in a number of the chapters in this book, as it can arise in a number of scenarios, including:

1. IBS
2. Functional dyspepsia
3. SIBO (see Chapter 24)
4. Carbohydrate intolerance, including FODMAPs
5. Excessive fibre intake.

Abnormal activity of the abdominal muscles

In addition to all of these factors, there is another mechanism that is thought to play a role in causing the symptoms of bloating and abdominal distension. This has the very long name of 'abdomino-phrenic dyssynergia'![78]

Normally when we eat, or have some gas in our gut, this causes a reflex relaxation of the diaphragm (it moves upwards) and a reflex tightening (contraction) of the muscles of the abdominal wall. This is called a normal 'viscerosomatic reflex' (VSR) and we show how this works in Figure 16. Many people with bloating have an abnormal VSR, such that the abdominal wall muscles relax and the diaphragm contracts (it moves downwards when this happens) and as a result their tummies stick out and become distended-looking.[77,79] This might be more of a problem for women who have been pregnant, as the abdominal wall muscles can become a little weakened during pregnancy, and in some women never quite return to normal.

How to beat bloating?

Just as there is no one, single cause, there is also no one-size-fits-all solution. But there are lots of things that can be done to help, which can add up to a very significant improvement. As with all digestive symptoms, it is important to mention any new symptoms to your GP/family doctor.

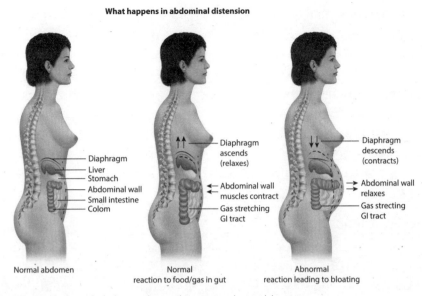

What happens in abdominal distension

Normal abdomen

Diaphragm
Liver
Stomach
Abdominal wall
Small intestine
Colom

Diaphragm ascends (relaxes)
Abdominal wall muscles contract
Gas stretching GI tract

Normal reaction to food/gas in gut

Diaphragm descends (contracts)
Abdominal wall relaxes
Gas strecting GI tract

Abnormal reaction leading to bloating

Figure 16 Abnormal abdominal muscle activity during bloating

1 Rule out other conditions

Depending on the symptoms, it may be worthwhile checking for SIBO. Many people wonder if they might have coeliac disease because they find that bread makes them bloated. It is important to ask your GP/family doctor to check you for coeliac if you notice this, but as we mentioned, there are a number of things in bread that can cause bloating, apart from gluten.

2 Diet

Diet is very important and seeing a dietitian can be extremely helpful. Carbohydrate foods, particularly fermentable carbohydrates, are the biggest culprits and it is generally a question of how much of these foods you eat rather than having to avoid them altogether. The balance of fibre in your diet is incredibly important – too little and you might have constipation, but too much can cause bloating. Fatty foods can also cause bloating – particularly in people with functional dyspepsia.

Some people also tend to swallow air when cigarette-smoking, chewing gum, or eating and swallowing too quickly. This is called aerophagia. Carbonated drinks such as sparkling water can also contribute to bloating and should be avoided.

3 Menstrual cycle

The menstrual cycle and time of the month can also be a factor for some women; many women experience more bloating in the week before their period. Hormonal factors cause us to retain fluid, but also affect the muscles in our gut, often causing the movement in the gut to slow down, causing constipation. If your menstrual cycle seems to play a big role, then it is important to track your gut symptoms and your cycle to pinpoint when your symptoms are most problematic. Paying particular attention to diet around these times can be very helpful.

4 Weight management

Weight gain of just seven pounds can also make us feel bloated. However, it is abdominal fat causing that feeling, not gas. Therefore, if you have a tendency to gain weight around the abdomen, losing some weight can ease the sensation of bloating.

5 Bowel habit/managing constipation

People with constipation are more likely to complain of bloating, than people who have a regular bowel pattern or diarrhoea. Therefore managing constipation, as we talked about in Chapter 14, is an important part of reducing bloating.

6 Probiotics

Various different combinations of probiotics have been investigated for their effects on bloating, and unfortunately there is no consistent evidence to support their use. We do not routinely recommend them.

7 Medications

Simple anti-spasmodic medications such as mebeverine or hyoscine are of benefit to some people. Substances such as activated charcoal and simethicone are available without prescription; some people find them helpful, but there is little evidence to support their use. A number of different medications that act on the gut–brain axis have been shown to be of benefit in certain subgroups of patients, and these are listed in Appendices 1 and 2.

8 Exercise

Working on core abdominal muscle strength and actively focusing on regular exercises to pull in and relax the abdominal muscles may help over time. For gut health in general we recommend regular exercise.

9 Complementary and alternative medicines

Some herbal and plant-based medications used in the management of functional dyspepsia may also be helpful in the management of bloating (Appendix 2).

How to manage bloating	
Do	**Don't**
Get the correct diagnosis if possible. Is it IBS or FD? Exclude organic conditions such as coeliac; possibly get checked for SIBO.	Drink carbonated drinks.
Avoid constipation. If you are constipated, try the measures advised to optimize the bowel habit.	Chew gum.
Assess your fibre intake. Aim to eat 20–35g fibre per day. If increasing your fibre intake, do it gradually by 5g per week.	Increase the amount of fibre in your diet quickly.
Limit your intake of highly fermentable carbohydrates.	Eat quickly.
Take regular exercise	Eat high fat meals.
Try exercises to improve your core strength. Many women have weakened abdominal wall muscles. Exercises to tone the abdominal wall, in conjunction with controlled breathing might be of benefit.	**Ignore abdominal weight gain.** This can add to the discomfort of bloating and add to the visible distention.
Consider medications. If your symptoms are not responding to first-line measures, discuss a trial of medication with your doctor.	**Despair!** Trying to deal with bloating is frustrating and upsetting. But small changes in a number of areas can amount to a big overall improvement.

Conclusion

Bloating is one of the most bothersome and frustrating digestive symptoms. Particularly frustrating is the fact that there is no quick fix. However, having a better understanding of how and why bloating occurs can help you deal with the problem in a more mindful way. Small changes in a number of areas can lead to a very significant overall improvement.

16

Management (treatment) of functional dyspepsia

A quick recap

As with IBS, the management of FD must be individualized and based on your symptoms, although there is some general advice that can be helpful to most people. The basic strategies are very similar to IBS, although there are some important differences. We must not forget that there is also a huge overlap between IBS and FD and that many patients may have both conditions (up to 50 per cent of them will have symptoms of both conditions), although one might predominate. This means that treatments may need to address both conditions.

Symptoms of functional dyspepsia: quick recap

- Post-prandial fullness
- Early satiety (feeling full after a small amount of food)
- Epigastric pain or burning
- Nausea
- Excessive belching
- Post-prandial bloating

Most of the symptoms occur after eating (post-prandial), although the pain in the upper abdomen may be more constant. If you get unpleasant symptoms whenever you eat, then it is not surprising that you might start to dread eating and to worry about these symptoms coming on, even before you eat anything. This is called 'symptom anticipation'. This leads to a whole additional level of anxiety, which in turn has a negative effect on the symptoms.

- **Personalized treatment:** The management of FD must be personalized. If you experience pain and burning, your solution may be different from someone who has post-prandial bloating as their main symptom.

- **Management, but not a cure:** As with IBS, there is no cure for FD but there are many, many things than can be done to help your symptoms.
- **Find the right clinician:** As we discussed for IBS, it's vital that you click with your doctor and that you feel that they are taking your concerns seriously.
- **Appropriate tests:** Make sure that other conditions have been ruled out.
- **Holistic approach:** Just like IBS, FD is a disorder of gut-brain interaction and also requires a holistic, step-wise approach that addresses every aspect of your wellbeing. To help with this, we recommend the same TEAMS approach as we showed you for IBS.

The Gut Experts' 4-Step FD Solution

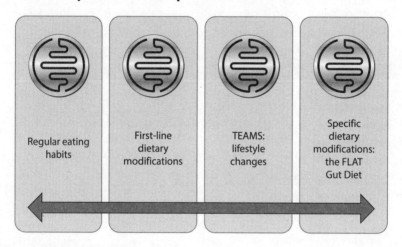

| Regular eating habits | First-line dietary modifications | TEAMS: lifestyle changes | Specific dietary modifications: the FLAT Gut Diet |

Step 1: Regular eating habits

This is particularly important with FD. If we consider that FD symptoms are often brought on by eating then it makes sense that how and what you eat could be triggers. 'How you eat' is almost as important as what you eat. Here's what we know:

- **Eat at regular mealtimes:** Irregular eating patterns are associated with worse FD symptoms.[80,81]

- **Smaller portions:** Eating less at one sitting and a little more often may be of value in FD ('little and often'). Spreading one's food out over three or four smaller meals, rather than two or three larger meals over the course of the day may be helpful. This does not mean that you should graze continuously all day; remember your house-keeping wave!
- **Eat slowly:** You should eat more slowly as fast eating is associated with worse symptoms.
- **Avoid cold fluids:** Very cold fluids can cause intense contractions of the stomach muscle, more marked in patients with FD. Therefore, if you have FD, you should avoid very cold/icy liquids.

Step 2: First-line dietary modifications for FD

Dietary management of functional dyspepsia
Dietary fat: High-fat foods, particularly saturated (animal) fats can worsen FD symptoms. A Mediterranean-style diet has been shown to be beneficial.
Dietary wheat: Wheat and gluten in the diet are associated with worsening of both post-prandial distress-type symptoms and of epigastric pain. Reducing wheat content in the diet may be helpful.
Caffeine: Caffeine can worsen epigastric pain symptoms.
Chilli and red peppers: These both contain capsaicin, which activates pain receptors in the stomach and can worsen FD symptoms. Interestingly, high daily intake of chilli can actually desensitize the pain receptors in the stomach.
Natural salicylates: Salicylates (present in aspirin) are present in many foods and can trigger FD symptoms. Common foods that contain salicylates include avocado, herbs, spices, capsicum, chilli, onion, tomato, aubergine, and olives.
Alcohol: Worsens FD symptoms, particularly epigastric pain and burning. Should be kept to a minimum.
Chewing gum: Can lead to swallowing of air while chewing and can worsen bloating.

Step 3: TEAMS: lifestyle measures

Skip to Chapter 23 for specific TEAMS advice and targets.

Exercise

We suggest following the same advice as we outlined for IBS.

Alcohol

One large study found that drinking more than 7 units of alcohol per week worsened FD symptoms, while another study found that over half of patients with FD reported that alcohol worsened their symptoms, with wine and beer being the most problematic drinks.[80,81] Therefore, our advice is that if you have FD you should keep your alcohol to a minimum.

Mental health

Stress and anxiety worsen FD symptoms and similar strategies as used in IBS can help. As we mentioned above, 'symptom anticipation' can be particularly problematic for patients with severe FD, and this anxiety often needs to be treated medically to try to 'break the cycle' of negativity. Cognitive behavioural therapy can also prove very helpful in helping reduce this food-related anxiety. Hypnotherapy and acupuncture may have a role particularly when used in conjunction with other therapies. Mindfulness and other forms of relaxation that help to reduce your stress are also beneficial.

Sleep

Just as with IBS, adequate sleep and good sleep hygiene are vitally important for people with FD.

Step 4: Specific dietary modification: the FLAT Gut Diet

Exclusion diets such as the low FODMAP diet can benefit people with FD, particularly those with post-prandial distress syndrome (PDS) (the bloating type of FD). Results of this type of dietary approach are less convincing for people with epigastric pain syndrome (EPS). In our practice, we find that the FLAT Gut Diet can be helpful for people with FD-PDS, but we do not use it routinely for people with FD-EPS. Therefore if you have FD-PDS and your symptoms are not adequately controlled by the first 3 steps we have just outlined, we would love for you to try the FLAT Gut Diet.

And remember... medications can be helpful

Medications

Some people, particularly those with FD-EPS, will require medications to reduce acid production in the stomach and to reduce hypersensitivity of nerves in the stomach. We often prescribe medications in conjunction with dietary and lifestyle measures from the outset, because we know from experience that diet and lifestyle alone will not control symptoms enough. There are many different medications available for the treatment of functional dyspepsia, and no single medication works in every person, so it can be a question of trial and error. These medications are generally only available on prescription.[82] Your GP or family doctor might prescribe some of these for you as a trial, but if you have more problematic or persistent symptoms you should be referred to a specialist for appropriate investigation and further treatment as necessary. We include some of the medications used for FD in Appendix 2.

Plant-based therapies (phytotherapy)

Some natural or plant-based substances are helpful for FD. Here are some of the plant-based therapies that have been studied:

Plant-based therapy	Effect
Ginger	Reduces nausea, post-prandial fullness and possibly early satiety
Peppermint oil & caraway oil (POCO)	Reduces epigastric pain, fullness, bloating and nausea
Iberogast A mixture of nine herbs: bitter candytuft, angelica root, caraway, fruit liquorice root, peppermint herb, balm leaf and chamomile flower	Reduces overall symptoms of FD, less pain and fullness
Almonds	Reduce overall FD symptoms

Total management of functional dyspepsia: the dos and don'ts

	Do	Don't
Eating habits	Eat slowly	
	Eat little and more often	
	Avoid cold fluids	
Diet	Eat lower-fat meals	Eat a large, heavy meal
	Reduce caffeine intake	Eat chilli, red peppers
	See a registered dietitian for formal advice	Drink more than 7 units of alcohol per week
Lifestyle	Get adequate sleep	
	Exercise regularly	
Stress and anxiety	Try mindfulness or other relaxation methods	Ignore on-going significant stress
	Seek professional help if unable to manage stress alone	
Medications	Speak to your doctor regarding a trial of medication	Take regular medication without appropriate medical advice
Herbal remedies	Try peppermint/caraway oil, or ginger or Iberogast	Try herbal remedies if you do not know exactly what they contain

Conclusion

Functional dyspepsia affects up to 20 per cent of people and, while the optimum therapy has not yet been clearly defined, if you have FD, there is a lot that can be done to help you. The best approach is through a combination of dietary and lifestyle changes, medications where appropriate, a trial perhaps of some plant-based therapies and trying measures to improve associated anxiety and other lifestyle factors. Just as with IBS, if you have problematic FD it is important to find a doctor/dietitian who has an interest in functional disorders and who will work with you to find the best possible solution for you.

Section 5

THE GUT EXPERTS' FLAT GUT DIETARY PROGRAMME

17

Introducing the Gut Experts' FLAT Gut Diet

Background

We're very excited to introduce you to, and get you started on the Gut Experts' FLAT Gut Diet. This dietary programme is based on our decades of experience involved in the clinical care of people just like you. We have diagnosed and treated tens of thousands of people with gut problems and we know, first hand, the toll that these conditions can take on your physical and mental health. It is both our passion, and our mission to share our knowledge with you, and to empower you to manage your gut symptoms and find a pathway to a happy, healthy gut.

In this section, we will explain how the FLAT Gut Diet works to improve the management of your IBS and gut symptoms. This diet can also be safely and successfully followed even if you don't have IBS, but simply want to follow a diet that results in very little bloating and a very comfortable gut.

'FLAT' is an acronym for the most important dietary and lifestyle components that trigger digestive symptoms in patients with IBS. We call these the FLAT Gut Factors. We spoke about these foods in Chapter 12, and why they have the potential to cause problems. But just to recap and focus your mind....

FLAT Gut Factors

F	Fibre
	Fructans
	Fructose
L	Lactose
A	Alliums
T	Total mind and body health

The FLAT Gut Diet: three steps

There are three steps in the FLAT Gut Diet.

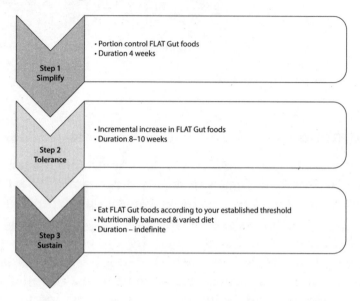

Step 1
Simplify
- Portion control FLAT Gut foods
- Duration 4 weeks

Step 2
Tolerance
- Incremental increase in FLAT Gut foods
- Duration 8–10 weeks

Step 3
Sustain
- Eat FLAT Gut foods according to your established threshold
- Nutritionally balanced & varied diet
- Duration – indefinite

STEP 1 – SIMPLIFY

In Step 1, you will be shown how to control the portions that you eat of the FLAT Gut food factors. We call this the 'Simplify' step and it lasts for four weeks. Apart from eating the FLAT Gut foods in measured amounts, you may continue eating other foods – remember, it does not need to be complicated. We do not want you to eliminate other foods completely as many of these foods have significant gut and overall health benefits.

We start with a low level of the FLAT Gut foods, which we know are generally well tolerated. You are allowed a certain amount of each of these foods per day. These foods are categorized by points, using our easy-to-follow points system.

Alliums (onions, garlic and the white part of leeks and scallions) are not allowed during this phase, because of their very potent effects on the gut of people with IBS. We also recommend that alcohol be kept to a minimum during this step. Legumes (beans and lentils) are also not allowed during this phase. While healthy for the gut bacteria, they can be very problematic for gut sufferers. Tolerance for these can be challenged at a later step.

STEP 2 – TOLERANCE

Building on what you have achieved in Step 1, during Step 2, you will be asked to gradually increase your daily intake of these FLAT Gut foods over 8–10 weeks. *You* will identify the food groups that *you* are most sensitive to. This is a very individualized phase of our dietary programme as you are trying to find *your optimal threshold* and tolerance level for these foods.

STEP 3 – SUSTAIN

In this maintenance phase, you will learn to sustain a nutritionally balanced, diverse diet while limiting your trigger foods. If you start to develop symptoms, you can always go back a step at any time to re-establish your equilibrium and get back on track.

THE FLAT GUT FOODS

Before we start Step 1, we want you to pare back your diet to relatively simple foods and to avoid any processed foods, which may contain hidden trigger ingredients such as onion or garlic powder.

We will run through each of the FLAT Gut food types, starting with fibre. Using our easy-to-follow points system, we have developed detailed tables to show you how much fibre, fructan, fructose and lactose are present in various commonly eaten foods. There are also some foods that we want you to avoid altogether in Step 1. These tables are all found in Chapter 20.

The fibre–fructan conundrum

Fructans are a particular type of fibre. We have told you that fibre is good for you, yet we are also telling you that fructans need to be limited. How on earth are you supposed to meet your daily fibre target and yet limit your fructans? Fortunately, there are lots of fibre-containing foods that do not contain or are low in fructans. These foods are also low fermentable, which is particularly important if you have IBS. So, the trick (and the challenge to be honest) is to get enough fibre in your diet, but in a low-fructan way.

- Legumes (peas, beans and lentils) are an excellent source of fibre, in the form of GOS. Unfortunately, GOS-containing foods are often problematic for people with IBS, and so we will consider these separately.

- Some foods can be considered 'free fibre', as they contain a low (or no) amount of fructan or other fermentable fibres. However, some of the fibre foods do carry fructan points and should be eaten as part of your daily fructan allowance. These foods are highlighted in the tables in Chapter 20.
- Fruits contain fibre but can also have a high fructose content, which needs to be limited each day. Suitable fruits can be chosen that contain a lower fructose content.

Fibre

We know the importance and benefits of fibre in your diet. Therefore, working out your daily fibre intake is your first essential step, if you have IBS or other gut problems. In The FLAT Gut Diet Plan one fibre point equates to one gram of fibre.

You may then realize that you are taking too much fibre and you need to reduce it. Depending on the severity of your symptoms, by reducing your fibre intake, you may see an immediate improvement in your symptoms. Some of you may be eating too little fibre and need to increase it in your daily diet. Choosing brown and wholegrain varieties rather than white for foods such as bread, pasta and rice, and eating potatoes and other vegetables with their skins on, will allow you to increase your daily fibre intake.

Your fibre intake should come from a *variety* of sources: soluble oat fibre, a small amount of wheat fibre and low gas by-product vegetables, salads and fruit fibres. This is what we mean by 'eating the rainbow'.

Vegetables and salads as a source of fibre

Vegetables are generally a good source of fibre, and add bulk and taste to our diet. Large amounts of many vegetables can be eaten freely, but smaller portions of certain vegetables are recommended as they contain fructans, fructose or polyols. Again, think *'variety'*.

Legumes as a source of fibre

Legumes (beans, peas, lentils, pulses and hummus) contain a range of indigestible fibres and they can also contain fructans, galacto-oligosaccharides (GOS) and fructose. Different cooking methods such as slow cooking, soaking, changing soaked or boiled water frequently, fermenting and sprouting can make legumes more digestible. Canned legumes in small quantities are often tolerated better than if cooked from dry.

However, despite these cooking methods, for many with IBS and functional dyspepsia, legumes can cause a lot of bloating, flatulence and pain. For this reason, in Step 1 – Simplify, we advise that you completely remove legumes from your diet. We realize that this can pose a significant challenge to vegans and vegetarians, but there are many other rich sources of suitable protein, including tofu, pea protein, tempeh, quorn, spirulina, nuts and seeds. When you progress to Step 2 – Tolerance, small amounts of legumes will be reintroduced in a controlled way.

Legume tips

If you really cannot live without legumes (ideally we would prefer that you wait until Step 2 – Tolerance) we suggest the following:

1. Eat legumes on a Friday evening after work, as opposed to Sunday evening, when you might have a work presentation to give early on Monday morning. In this way, side-effects (bloating and flatulence) from these foods can be managed a lot more easily if you are at home rather than at work the next day!
2. Tinned legumes are better tolerated than the dry varieties. Chickpeas and mixed beans in small quantities are the easiest to digest.
3. Alpha-galactosidase is a digestive enzyme that breaks down the carbohydrates in legumes into simpler sugars, thus making them easier to digest. This enzyme can be taken in supplement form, and may aid the digestion of legumes and reduce their side effects.

Fructans

We discussed fructans in Chapter 12. For many people, wheat is one of the main sources of fructans in the diet. Rye also has a very high fructan content – higher than that in wheat-based foods. As a result, we exclude rye in Step 1 of the diet.

Vegetables can also contain fructans, and we have included those vegetables that are best avoided in the list above.[36,83] Because alliums have a very high fructan content and are extremely potent,[84] we consider these vegetables as a separate entity in our dietary approach.

We do not want you to cut out fructans from your diet, because of their health benefits.[85] We just advise you to cut down the quantity, according to our specially devised Fructan Points System.

The fructan foods allowed here are bread, cereal, pasta, noodles and the biscuit group. The FLAT Gut Diet is not a gluten-free diet. We recommend gluten-free products only for those with a definite diagnosis of coeliac disease. While gluten-free products are generally suitable, they contain other ingredients, which, if taken in large quantities, may also cause digestive complaints and weight gain.

Inulin is a fructan that is used as an ingredient in many foods such as cereals, high-protein bars, health bars, muesli bars and a range of 'healthy' food products. While it doesn't affect most individuals, it can cause significant symptoms if you have IBS, such as wind, bloating and loose bowel motions.

So, when it comes to fructans, there are some grains and cereals naturally low in fructans that are freely allowed. There are other foods high in fructans that we recommend that you avoid altogether and some that you are allowed to eat according to our points system.

Bread tips

- Rye sourdough is high in fructans and can be problematic for the gut, therefore is best avoided.
- 100 per cent spelt sourdough that goes through the full fermentation process reduces fructan content. This bread is tolerated in larger quantities – e.g. two to three slices per sitting. Baking processes vary across manufacturers, so if fermentation time is shorter, there may be no benefits over other breads.
- Spelt flour is from an ancient wheat grain, and may have a lower fructan content, particularly if sieved. Spelt can be easier to digest than wheat, and is generally tolerated better. We have used 100 per cent spelt flour in many of our recipes.
- However, spelt flakes and spelt pasta are high in fructans and best avoided.

'Let's talk bread, pasta and pizza'

- Who does not love fresh bread and a lovely big bowl of pasta or fresh pizza? In general, our consumption of these foods has increased significantly.
- We know that for many people, it is the fructan part of the bread rather than gluten that causes IBS symptoms. Wheat-containing foods contain moderate amounts of fructans, but our consumption of these foods can be considerable, leading to diets high in fructans.
- It is often the portion sizes of these foods that causes problems, and smaller portions are usually well tolerated.
- The 'fructan' content of breads varies according to the amount of wheat flour, the milling process and whether other flours such as maize flour, are added. It is also dependent on the manufacturing process and cooking techniques.
- Many breads are baked from part-baked and frozen products in a central production unit. Freezing the bread changes the starch to make it resistant starch. This makes it harder to digest and allows it to be fermented by the gut bacteria, leading to more bloating.
- In the USA, wheat breads and baked products can contain high-fructose corn syrups, which can be problematic for some people.
- Many gluten-free products have similar fructan contents, and in some cases, more fructans than normal breads. Many factors may be at play here including processing techniques and added ingredients, for example maize flour.

Take-home message If wheat-based foods cause bloating and other gut symptoms, smaller portions are usually well tolerated and you may simply need to find your tolerance level.

Fructose

Fructose malabsorption is found in approximately one third of the population. Everyone has a fructose threshold: remember 'Alice's story' – she was unknowingly eating too much fructose. Fructose is poorly absorbed, particularly when there is more fructose than glucose in the food or it has a high fructose load per serving.[83]

High fructose-containing foods are commonly poorly tolerated. These foods include honey, apples, blackberries and watermelon, and should be avoided altogether. The fructose points table in Chapter 20 shows you the fructose points of allowed fruits and the ones to avoid completely.

Spread your fructose intake over the course of the day, leaving several hours in between eating fruit, so the fructose load at each sitting is well tolerated.

Fructose (fruit) load can be a problem when taken together with lactose (yoghurt) in large portions. Ideally in Step 1 of the FLAT Gut Diet, we advise eating fructose at a different time to lactose, or to take just a very small amount of fruit to flavour, such as two chopped strawberries added to a plain yoghurt.

Lactose

With the exception of those individuals who have a cow's milk protein allergy (which is rare in adults), avoidance of dairy products is not advised on the FLAT Gut Diet. As we saw with Michelle's story, a trial period of dairy avoidance can sometimes end up becoming years of avoidance, with long lasting consequences. We advise reducing the intake of foods high in lactose to smaller quantities, until you find your tolerance level. Even people with diagnosed lactose intolerance can tolerate daily lactose amounts of 12–15g per day.

In our points system for lactose, 4–5g lactose = 1 point. We recommend that you *do* consume dairy each day to ensure that you get adequate calcium in your diet.

If you are a vegan and do not consume dairy products, you can still follow this diet. Make sure that you are getting adequate calcium from other sources, such as fortified non-dairy products. You should also consider a calcium supplement.

Common lactose pitfalls

- Intake of too much lactose in one sitting has increasingly become a problem with the introduction of 500g pots of yoghurt. This is important to remember. It is so easy to eat over half of it in one sitting. A healthy snack, but the lactose content may be too much in one go for your gut. Choose smaller pot portions of yoghurt such as 100/125g.
- Yoghurt eaten with fruit (fresh and dried), muesli, nuts and seeds can contain a powerful combination of lactose, fructose, fructans and fibre. This overload commonly triggers uncomfortable gut symptoms, particularly if you have IBS.

- Large quantities of lactose (yoghurt) and fructose (fruits), drizzled with honey (fructose) taken for breakfast are a common cause of bloating and wind.
- Consume fruit and yoghurt separately, and we advise taking your yoghurt in the afternoon or evening time. If taking yogurt and fruit together only use a few permitted berries (from the fructose table) to flavour.

'Let's talk cheese'

- One matchbox-size portion of cheese provides about 220mg of calcium, approximately a third of your daily allowance.
- Aged cheeses are generally harder cheeses and contain minimal lactose. Examples include Brie, blue vein, Camembert, Cheddar, Emmenthal, Gruyere, Parmesan, Pecorino, soy, Swiss. Hard cheeses contain none or only trace amounts of lactose and are a great source of calcium. We recommend including these cheeses in your daily diet and they are well tolerated as a snack.
- Processed cheeses and spreads are melted natural cheeses. Milk or whey is often added for flavour, and so they can have a higher amount of lactose.

Lactose tips

- Read cheese labels – lactose is the main carbohydrate in cheese. A simple guide is to read the carbohydrate content on the label; low numbers tell us there are low amounts of lactose in the cheese. For example, Emmental has 0g carbohydrate per 100g and so has no lactose. It is therefore a great source of calcium without causing gut symptoms.
- Lactose may be tolerated better if consumed with other foods rather than on its own in between meals.
- There are lots of dairy foods that are low in lactose, such as lactose-free milk and many cheeses, which can be eaten freely and also boost your calcium intake.

Alliums

Alliums such as onions, leeks and garlic, are inulin-type fructans, and induce very potent gut symptoms. They are much stronger than wheat grains and cereals in causing gut symptoms. Because of this, we consider them in a category of their own, separately from other fructan-containing foods. Many years ago, they were not given as much consideration because they were only used in small quantities. However, the consumption of alliums and the quantities used in cooking have increased significantly in recent years, and can be troublesome if you have IBS or bloating.

The green part or leaf part of leeks, spring onions and chives do not cause digestive complaints, and are always permitted on the FLAT Gut Diet.

Common pitfalls with onions and garlic

- Onion and garlic are often used in products for flavouring. Avoid all crisps, dressings, gravies, ready-made meals, sauces, soups and stock cubes containing onion and garlic.
- It is nearly impossible to avoid these foods completely as they are in so many manufactured products. So, we can only do our best to minimize our intake, by not adding extra when preparing and cooking our food or meals.
- When reading labels, the ingredients are listed by weight, therefore the first ingredient is the most abundant ingredient in the food product and the last one is the least. However, onion and garlic powders weigh very little but can still have a very strong effect on the bowel.

Cheeky garlic cheat

If you cannot live without garlic (!) you can try commercially prepared garlic-infused oil or rub a clove of garlic around the pan, before adding other ingredients. This can give some flavour and aroma without inducing symptoms.

Pickled garlic can be tolerated in a very small amount, to garnish and flavour a meal.

Cheeky onion treat

Pickled onions are tolerated in a small quantity to garnish and flavour a meal. Check the label to ensure there is no fructose used in the preserving liquid.

Cooking without onions and garlic

At first it may seem unnatural to cook without onion or garlic, but it's easy to get used to. Your food doesn't have to be boring. There are lots of ways to flavour food with herbs, spices, citrus fruits, sauces and condiments. In Chapter 20, we include a table showing lots of flavourings that are allowed in the FLAT Gut Diet, giving you plenty of other options to add flavour and interest to your food.

Protein foods

- All animal-based protein foods (meat, fish, poultry and eggs) are very well tolerated by the digestive system. A diet very high in protein may lead to protein fermentation and some malodorous flatulence. However, this generally does not occur with normal daily quantities of animal proteins.
- Dairy foods are also animal-based proteins, but because many contain lactose, we consider these under the 'Lactose' section.
- Many plant-based proteins also contain fibre, fructans and GOS as we explained earlier when we discussed legumes. For this reason, we recommend avoiding all legumes in Step 1. They can be introduced in small amounts later in Step 2.
- Many plant-based proteins, including tempeh, tofu, quorn, nuts and seeds are suitable to eat in all phases of the diet.
- Nut and seed proteins are suitable in a small handful once per day. Nut butters are also well tolerated.
- There are only two nuts/nut butters that you need to avoid completely; these are cashew and pistachio nuts. They are high in fermentable components (GOS).

Fats and oils

- In general, people tolerate most fats and oils when on the FLAT Gut Diet.
- Fats are an important part of a healthy diet, essential for skin and heart health.
- Mono- and poly-unsaturated fats are the best choice for overall and gut health, and these include the following oils: olive, avocado, almond, sesame, rapeseed, walnut, sunflower, peanut and rice bran.

- Saturated fats such as butter, margarine, ghee, mayonnaise, palm oil, coconut oil and canola oil are also well tolerated but should be eaten in moderation.

Beverages and alcohol

- **Water** is recommended as the most natural choice of fluid for hydration.
- **Alcohol** has many potential adverse effects on the gut and you should keep your alcohol intake to a minimum in Step 1 of the FLAT Gut Diet. In Step 2 we recommend a maximum of 2 units per day, but ideally less and not every day.
- **Fruit juices** are high in fructose and can cause gut symptoms. Juices such as apple and pear juices, as well as smoothies, are made with a lot of high fructose fruits.
- **Carbonated drinks** should be taken in moderation as large quantities lead to gas and wind.
- **Caffeine** acts like a double edged-sword; on the one hand, it can help alleviate constipation, but on the other, it can speed bowel transit too much, causing diarrhoea. Caffeine may also worsen symptoms of functional dyspepsia.
- **Isotonic sports drinks** often have some added sodium and glucose to aid absorption and are easily absorbed and suitable for hydration during longer duration exercise or high intensity exercise, where fluid loss may be high.
- **Hypertonic sports drinks** contain a high concentration of sugar and can cause diarrhoea.

Sweeteners and polyols

- **Honey and agave nectar** although natural, are high in fructose and can be poorly absorbed by some people. It seems like a paradox, but ordinary table sugar (sucrose) is easily absorbed in the bowel. This means that desserts, like sweet meringues, are easily absorbed, whereas stone fruits drizzled with honey are poorly absorbed by those with IBS.
- **Fructose corn syrups** and ingredients added to foods can play a big role in causing gut symptoms. These are widely used as sweeteners in the USA and Canada.

- **Sugar-free gum** contains polyol sweeteners. One or two pieces may be tolerated in a day, but a packet a day will likely produce symptoms. Chewing gum by the very nature of the chewing, leads to aerophagia (swallowing of air). This may play a part in contributing to bloating.
- **Polyols** are not tolerated in large quantities in the daily diet as they are very poorly absorbed in the gut. Foods containing sorbitol and mannitol are not recommended in Step 1 of the FLAT Gut Diet.
- Polyols are also found in some fruits such as apple, apricot, avocado, blackberries, cherry, nectarine, peach, pear and plum. These are to be avoided in Step 1.
- Certain vegetables such as cauliflower and mushroom are high in polyols and are to be avoided.
- Coconut dessicated is a polyol and can be tolerated in small quantities.

Summary: ready to start?

Phew! You have done a lot of reading up to this point, but the good news is that now you're ready to get started on the FLAT Gut Diet. You will need to refer to the food points and portions tables in Chapter 20. These show you:

- the foods that are allowed freely;
- foods that are allowed but need to be portion controlled;
- foods that you should avoid completely in Step 1.

Use the easy-to-follow points system for fibre, fructans, fructose and lactose, as you set out on your 12-week journey. At the end of this period, we hope that you will have your symptoms under better control than ever before, and that you will also have a good understanding of your threshold levels for the main FLAT Gut food factors.

We have talked in detail about all the FLAT Gut food factors, but don't forget that the 'T' stands for 'Total mind and body health'. If you want to skip forward to Chapter 23 you can also put your TEAMS plan in place to go hand in hand with your diet.

Now, let's get started on the diet.

18

Before you start: your shopping list and stock items

Shopping list for the FLAT Gut Diet 7-Day Meal Plan

The following list details the fresh foods you will need to purchase in order to follow the meal plan for 7 days, including approximate quantities to make meals to recipe portions (typically two to four portions). It also includes a wide variety of store cupboard items (many of which you'll probably find in your cupboard at home and won't need to buy), that you will need throughout the FLAT Gut Diet.

Fresh produce
Fruit

- ☐ Bananas – 6
- ☐ Blueberries – 2 punnets (70–80)
- ☐ Grapes – 1 punnet (30)
- ☐ Kiwi – 2
- ☐ Lemons – 2
- ☐ Lime – 1
- ☐ Mandarin/Clementine oranges – 4
- ☐ Raspberries or strawberries – 2 punnets (30)
- ☐ Strawberries – 320g

Vegetables & salad

- ☐ Asian green vegetables, such as baby pak choi/bok choy and choy sum – 400g
- ☐ Aubergines – 2 large
- ☐ Avocado – 2
- ☐ Baby cucumbers – 200g
- ☐ Baby gem lettuces – 4
- ☐ Baby salad leaves – 1 bag
- ☐ Baby spinach leaves – 1 bag
- ☐ Beansprouts – 25g
- ☐ Broccoli, tender stem – 200g
- ☐ Carrots – 2kg
- ☐ Celery sticks – 7
- ☐ Cherry tomatoes – 750g
- ☐ Courgettes – 2 small
- ☐ Cucumber – 1
- ☐ Green beans, extra fine – 150g
- ☐ Fennel bulb – 1
- ☐ Leeks (green part) – 5
- ☐ Mangetout – 80g
- ☐ Parsnips – 2 small
- ☐ Red cabbage – 150g
- ☐ Red chillies – 4
- ☐ Red pepper – 1
- ☐ Root ginger – 10cm chunk
- ☐ Spring onions (green part) – 12
- ☐ Sugar snap peas – 80g
- ☐ Swede – 1 small
- ☐ Watercress – handful
- ☐ Yellow pepper – 1

Staple starches

- ☐ 100% spelt sourdough bread loaf or 4–8 slices wholemeal bread/soda bread
- ☐ Potatoes, baby – 500g
- ☐ Potatoes – 2 medium
- ☐ Pasta – 200g fresh egg pasta or 300g buckwheat pasta

Fridge/freezer
Dairy

- ☐ Butter – 250g
- ☐ Buttermilk – 400ml
- ☐ Cheese, Cashel or Crozier blue – 50g
- ☐ Cheese, feta – 200g
- ☐ Cheese, goat's – 2 x 100g individual
- ☐ Cheese, parmesan – 2 tablespoons
- ☐ Cheese – variety of suitable options 250g
- ☐ Cream – 500ml
- ☐ Crème fraiche – 500g
- ☐ Mayonnaise – 2 tablespoons
- ☐ Milk, whole or low-fat – 660ml or lactose-free – 800ml or hazelnut, almond, rice, hemp or other calcium-enriched milk alternative – 850ml
- ☐ Sour cream – 1 tablespoon
- ☐ Yoghurt – natural, strawberry or raspberry – 400g
- ☐ Yoghurt, Greek – 150g

Meat, fish, poultry, eggs & tofu

- ☐ Chicken fillets – 800g
- ☐ Chicken – 1 large
- ☐ Eggs – 16 large
- ☐ Beef, lean minced – 400g
- ☐ Prawns – 75g
- ☐ Salmon fillets – 4 x 150g

- [ ] Sirloin steaks – 4 x 175g
- [ ] Smoked salmon ribbons – 300g
- [ ] Smoked streaky bacon – 150g
- [ ] Tofu – 400g

Fresh herbs – 1 packet/plant of each

- [ ] Basil, chives, coriander, flat-leaf parsley, oregano, thyme

Store cupboard

Grains and flours

- [ ] Cornflour
- [ ] Porridge oats
- [ ] Wheat germ
- [ ] White flour and/or 100% spelt white flour
- [ ] Wholemeal flour and/or 100% spelt wholemeal flour

Rice, noodles, crackers

- [ ] Crackers, cream/salted
- [ ] Rice noodles
- [ ] Rice cakes
- [ ] Rice, Thai jasmine
- [ ] Rice, wholegrain

Dried fruit, seeds, nuts and nut butters

- [ ] Dried fruit: banana chips, cranberries, raisins, shredded/desiccated coconut
- [ ] Seeds: black sesame, chia, linseeds, mixed, pumpkin, sunflower, white sesame
- [ ] Nuts: almonds, hazelnuts, macadamia, pecan, walnuts
- [ ] Nut butters: almond, peanut

Baking ingredients

- ☐ Baking powder
- ☐ Bicarbonate of soda
- ☐ Sugar: caster
- ☐ Syrup: golden
- ☐ Syrup: maple
- ☐ Vanilla extract (bottle)

19

Step 1 – 'Simplify' and 7-day meal plan

Getting started: your food diary

We've included a sample food, fluid and symptom diary for you to complete each day of Step 1 (Appendix 3). You should record all your symptoms, your dietary intake and all the other relevant factors we have included. This will help you to track your progress.

Step 1 – Simplify

The aim of the 4-week Simplify phase of the diet is to start with low levels of the FLAT Gut foods, which we know to be well tolerated. This should significantly reduce your symptoms of bloating, abdominal distension, pain and flatulence, while ensuring adequate and varied nutrition. We are starting at low levels of these foods, but aim to work them up over time, by finding your 'tolerance level' in Step 2 – the Tolerance phase.

The FLAT Gut Diet: Step 1 – Simplify		
Fibre	**20**	Points per day
Fructans	**2–3**	Points per day
Fructose	**2–3**	Points per day
Lactose	**2–3**	Points per day
Alliums	**0**	Points per day
Total		Wellbeing – mind and body

Practicality & flexibility

Following any diet requires planning, particularly if you work outside the home. You may need to bring lunch and snacks with you for during the day. We fully realize that life is not always predictable and some days it may not be possible to get it 100 per cent right. But you don't need to be perfect! While you do not have to reach the total number of points for each FLAT Gut food in a given day, we do recommend that you pay particular attention to your fibre content. If you do not reach 20 points per day, you may tend towards constipation. On the other hand, some of you may be eating far too much fibre and need to pare it back to start with, to find your tolerance level. The Lactose points are also very important, to ensure that you are meeting your daily calcium requirements. In reality, the total fructans and fructose in any one day are less important – so long as you don't go over the recommended points.

Sample 7-Day Diet Plan for Step 1

- **Getting the points right:** In this sample 7-day eating plan, we include meals that are suitable for the Step 1 – Simplify phase of the diet. This means that each day contains approximately 20 fibre points, 2–3 points for fructan and fructose, and 2–3 lactose points.
- **Easy prep:** The meals are simple to prepare and most of the lunches are easily brought to work in a lunchbox.
- **Don't mix your sugars:** As we advised, we would like you to eat your lactose and fructose separately where possible. The small amounts of fruit that we have included with milk at breakfast, are low-fructose fruits, and should not cause any problems.
- **Just eat what you need:** You don't have to eat everything on the diet! The amount of food may be too much for some people. We advise you to respond to your own hunger levels. A simple plant-based meal/salad at lunchtime may be enough.
- **Snacking:** You may not need all the snacks.

Step 1 – Simplify: FLAT Gut Diet 7-day Meal Plan

Meal time	Monday	Tuesday	Wednesday	Thursday	Friday	Saturday	Sunday	Notes
Breakfast	Tailor-made porridge	FLAT Gut muesli Milk 100ml (½ glass)	FLAT Gut breakfast bar 150mls Kefir	FLAT Gut oat shake	FLAT Gut muesli Milk 100ml	Blueberry & banana power pancakes	Tailor-made porridge	
Mid-morning	Banana 1 Cheddar cheese 30g Crackers 2	Peanut butter Rice cakes 2	Crackers wholemeal 2 with Brie cheese 40g	Rice cakes 2 Peanut butter 45g	Mandarin 2 small	Fresh mint tea Walnuts 15g	Suitable herbal tea from list	
Light meal	Poached eggs with roasted vine tomatoes and smashed avocado on wholemeal bread	Mediterranean Greek salad with baked feta Cream crackers 4	Tuna mayo gherkin salad 1 portion Bread wholemeal 2 slices	Spiced carrot & ginger soup Ham and cheese on FLAT Gut brown soda bread	Oriental layered salad in a jar	Prawn & red cabbage slaw on 2 slices 100% spelt sourdough bread or FLAT Gut brown soda bread 60g	Smoked salmon, scrambled egg and FLAT Gut brown soda bread 60g	

Step 1 – Simplify: FLAT Gut Diet 7-day Meal Plan								
Meal time	Monday	Tuesday	Wednesday	Thursday	Friday	Saturday	Sunday	Notes
Afternoon	Mandarin 2 small	Grapes 10 Cheddar cheese 30g	Grapes 10	Banana Cheese 30g	Yoghurt 100g natural/ strawberry/ raspberry	Kiwi 2 Biscuits plain 2	Banana 1 Biscuits plain 2	
Main meal	Chinese-style chicken with Asian greens	Sesame crusted salmon & pickled vegetable salad Baby boiled potatoes	Sweet and sour tofu (or poultry) stir-fry vegetables Rice noodles	Keralan chicken curry with green beans and brown rice Mixed leaf salad	Spaghetti Bolognese	Roast aubergine with cherry tomatoes and goat's cheese	Sirloin steak and Cashel blue butter and roasted vegetables or fish dish (Bonus Recipes)	
Supper	Yoghurt 100g natural/ strawberry/ raspberry 15g chopped nuts	Greek yoghurt 150g Walnuts 30g chopped	Yoghurt 100g natural/ strawberry/ raspberry	Yoghurt 100g natural/ straw-berry/ raspberry	IBS-friendly cheese board with white crackers & grapes 10	Eton Mess	Chocolate brownies	

The points system in action

The following tables show you a daily breakdown of the FLAT Gut food points for the 7-day meal plan, so that you can see how the points system works.

Meal time	Food	Fibre points	Fructan points	Fructose points	Lactose points
Monday FLAT Gut Diet points system sample week's diet					
Breakfast	Tailor-made porridge oats	5			
	Fruit/milk with recipe			1	1
Mid-morning	Banana	1		1	
	Cheese 30g				0
	Crackers wholemeal 2	0.5	1		
Lunch	Poached eggs with roasted tomatoes and avocado	1.5			
	Wholemeal bread 2	4	2		
Afternoon	Mandarins 2 small	1		1	
Evening meal	Chinese-style chicken with Asian greens	4			
	Brown rice	2			
Supper	Yoghurt 100g natural/straw-berry/raspberry flavour 15g chopped nuts	1			1
Total		20	3	3	2

Tuesday FLAT Gut Diet points system sample week's diet					
Meal time	Food	Fibre points	Fructan points	Fructose points	Lactose points
Breakfast	FLAT Gut muesli	5			
	Fruit/milk with recipe			1	1
Mid-morning	Rice cakes 2		1		
	Peanut butter	1			
Lunch	Med. Greek salad with feta	4			
	Cream crackers white 4		2		
Afternoon	Grapes 10	1		1	
	Cheddar cheese 30g				
Evening meal	Sesame crusted salmon & pickled vegetable salad	4			
	Baby boiled potatoes	3			
Supper	Yoghurt 100g natural				1
	Walnuts chopped 30g	2			
Total		20	3	2	2

Wednesday FLAT Gut Diet points system sample week's diet					
Meal time	Food	Fibre points	Fructan points	Fructose points	Lactose points
Breakfast	FLAT Gut breakfast bar 100g	6			
	Kefir 150mls				1
Mid-morning	Crackers wholemeal 2	0.5	1		
	Cheese Brie 40g				0
Lunch	Tuna mayo, gherkin salad	2			
	Bread wholemeal 2 slices	4	2		
Afternoon	Grapes 10	1		1	
Evening meal	Sweet and sour tofu (or poultry) stir-fry vegetables & pineapple	7		1	
	Rice noodles				
Supper	Greek yoghurt 130g				1
Total		20.5	3	2	2

Thursday FLAT Gut Diet points system sample week's diet					
Meal time	Food	Fibre points	Fructan points	Fructose points	Lactose points
Breakfast	FLAT Gut oat shake	4			
	Milk/fruit in shake			1	1
Mid-morning	Rice cakes 2		1		
	Peanut butter	2			
Lunch	Spiced carrot & ginger soup	5			
	Ham & cheese				
	FLAT Gut brown soda bread 2	3	2		
Afternoon	Banana	1		1	
	Cheese 30g				0
Evening meal	Keralan chicken curry and green beans	2			
	Mixed leaf salad	1.5			
	Brown rice	2			
Supper	Yoghurt 100g natural/strawberry/ raspberry flavour				1
Total		20.5	3	2	2

Friday FLAT Gut Diet points system sample week's diet					
Meal time	Food	Fibre points	Fructan points	Fructose points	Lactose points
Breakfast	FLAT Gut muesli	5			
	Fruit/milk with recipe			1	1
Mid-morning	Mandarins 2 small	1		1	
Lunch	Oriental layered salad in a jar	8			
Afternoon	Yoghurt 100g natural/strawberry/ raspberry flavour				1
Evening meal	Spaghetti Bolognese	5	2		
Supper	IBS-friendly cheese board		1		0
	White crackers 2/3				
	grapes 10	1		1	
Total		20	3	3	2

Saturday FLAT Gut Diet points system sample week's diet					
Meal time	Food	Fibre points	Fructan points	Fructose points	Lactose points
Breakfast	Pancakes blueberry & banana power	3		1	
	Milk 1/2 glass 100ml				1
Mid-morning	Fresh mint tea				
	Walnuts 15g	1			
Lunch	Prawn & red cabbage slaw on 100% spelt sourdough bread	3 2	0 points if fully fermented		
Afternoon	Kiwi x 2	2		1	
	Biscuits plain x 2		1		
Evening meal	Roast aubergine with cherry tomatoes and goat's cheese	5			1
Supper	Eton Mess, strawberries, dark chocolate	2 2		1	1
Total		20	1	3	2

Sunday FLAT Gut Diet points system sample week's diet					
Meal time	Food	Fibre points	Fructan points	Fructose points	Lactose points
Breakfast	Tailor-made porridge	5		1	1
Mid-morning	Suitable herbal tea				
Lunch	Smoked salmon, scrambled egg and FLAT Gut brown soda spelt bread	3	2		
Afternoon	Banana	1		1	
	Plain biscuits x 2		1		
Evening meal	Fish dish (bonus recipe) or Sirloin steak and Cashel blue butter and roasted vegetables and potatoes	8			(1)
Supper	Chocolate brownies and crème fraiche	3			0.5
Total		20	3	2	1.5

Portion sizes and points system

- Recommended serving sizes are a guideline and not set in stone.
- Fruit sizes can vary.

Step 1 Weeks 2, 3 & 4

- We've included a one-week sample menu for the diet. You can repeat some of these meals in the remaining three weeks or use some of our bonus recipes.
- However, ideally, you'll learn to use the food points and portion tables to start to prepare and cook meals for yourself using ingredients that you like.
- We'd like you to prepare meals, working within your points allowance, so that eventually it becomes second nature.
- At the end of the 4 weeks of the Simplify phase, you should experience significant relief from your IBS symptoms, and will feel more in control.

Next steps

- Chapter 20 shows you the FLAT Gut Diet points and portion tables.
- Chapter 21 contains all the recipes for the 7-Day meal plan, some bonus recipes, snacks and beverages.
- In Chapter 22, you will see how to start Step 2 – Tolerance, to find your tolerance levels for the FLAT Gut foods.

20

The FLAT Gut Diet Plan: points and portions

FIBRE POINTS AND PORTIONS

FIBRE TABLE 1: Fibre content of commonly eaten foods		
Fibre foods	Portion size	Fibre (points per portion) The FLAT Gut Diet Step 1 – Simplify
Cereals and grains		**Start at 20 points per day**
FLAT Gut muesli	50g	5
FLAT Gut breakfast bar	1 bar (100g)	6
Oatabix	2 biscuits (48g)	4
Oatabix flakes/Quinoa flakes	40g	3
Tailor-made FLAT Gut porridge	1 serving	5
Porridge oats	40g	3
FLAT Gut oat shake	180ml	4
Rice bran/Oat bran	15g	3
Buckwheat flakes	40g	2
Cornflakes/Rice Krispies	40g	1
Breads, crackers, pitta, wraps		
Wholemeal bread*	2 slices (30g per slice)	4
Brown soda bread* (the FLAT Gut Diet recipe)	1 thick loaf slice (60g)	3
Brown spelt soda bread*	2 small slices (35g per slice)	2–4
Wholemeal bagel*/Pitta*	1 medium (60g)	4
Wholemeal wrap*	1 medium (60g)	3

(Continued)

FIBRE TABLE 1: Fibre content of commonly eaten foods		
Porridge bread	1 thick loaf slice (60g)	3
100% Sourdough spelt bread***	2 small slices (60g) or 1 large (60g)	1–3
Wholemeal crackers*	4	1
Nuts and seeds****		
Chia seeds	1 tablespoon (15g)	6
Linseed/flaxseed	1 tablespoon (15g)	4
Almonds	1 tablespoon (15g)	2
Brazil nuts, peanuts, sunflower seeds, walnuts	1 tablespoon (15g)	1
Almond butter	1 tablespoon (20g)	3
Peanut butter	1 tablespoon (20g)	1
Rice, pasta, potato		
Wholewheat pasta*	60g dry weight	6
Pasta – buckwheat	75g dry weight	6
Quinoa	75g dry weight	5
Potato (flesh & skin)	2 medium (150g)	4
Pasta – wheat pasta (white)*	60g dry weight	3
Pasta – quinoa	75g dry weight	3
Potato (flesh only)	2 medium (150g)	3
Sweet potato	70g cooked weight	2
Rice – brown	75g dry weight	2
Rice – white long grain/ basmati	75g dry weight	1
Fruit/vegetables/salad		**Average fibre points**
Fruit**	80g	1–2
Vegetables**	80g	2
Salad**	80g	1.5

*These foods need to be portion controlled due to their high fructan content. Please refer to fructan tables 1 and 2 for further details.

**Fibre content of these foods has been averaged.

***This is well tolerated but can be quite difficult to source. It may be best to try to make your own at home, which can also be a challenge! Fibre content will vary depending on brand and size of slices.

****Pistachio and cashew nuts & butters should be avoided in Step 1 due to the high content of fermentable components (GOS).

FIBRE TABLE 2: VEGETABLE AND SALAD TABLE	
Suitable vegetables and salads to be counted as fibre points	
80g vegetables = 2 fibre points, 80g salad = 1.5 fibre	
Alfalfa	Ginger
Aubergine/eggplant	Kale
Bamboo shoots	Lettuce (butter, cos, iceberg, lamb's, radicchio, red coral, romaine)
Beans, French long green	Mushroom, oyster
Beansprouts	Okra
Bell pepper (green, red, yellow, orange)	Olives
Bok choy	Pak choi
Cabbage (white and red)	Parsnip
	Pickled gherkins
Carrot	Plantain
Celeriac	Potato
Chard/Swiss chard	Radish
Chicory leaves	Rocket
Chilli, red and green	Seaweed/nori
Chives	Spinach
Choy sum	Swede
Courgette/zucchini	Tomato
Cucumber	Turnip
Endive beans	Water chestnuts
Fennel leaves	Yam
Vegetables and salads to be portion controlled – one small portion per day	
40g = 1 fibre point, 20g = 0.5 fibre point	
40g allowance per meal	**20g allowance per meal**
Artichoke hearts	Beetroot
Asparagus	Celery
Avocado	Mangetout
Broccoli	Mushroom, porcini
Brussels sprouts	Peas
Butternut squash	Sugar snap peas
Corn on the cob	Sweetcorn
Fennel bulb	

(Continued)

FIBRE TABLE 2: VEGETABLE AND SALAD TABLE	
Pumpkin	
Savoy cabbage	
Sweet potato	
Vegetables to be avoided due to poorly absorbed polyol content	
Cauliflower	Mushrooms
Chicory root	Sauerkraut, fermented

FIBRE TABLE 3: LEGUMES
The FLAT Gut Diet Step 1 – Simplify
Avoid all legumes (beans, lentils, pulses and hummus), cashew and pistachio butters and nuts. These foods, while very healthy, have a high GOS content.

FRUCTAN POINTS AND PORTIONS

FRUCTAN TABLE 1: Portion-controlled foods POINTS CHART		
Fructan-containing carbohydrate foods	**Portion size**	**Fructan points The FLAT Gut Diet Step 1 – Simplify 2–3 points daily**
Breads, crackers, rice, corn cakes		
Wholemeal/white	2 slices (30g per slice)	2
Wholemeal wrap/pitta/bagel	1 small (60g)	2
Wholegrain/multigrain	1.5 slices (45g)	2
FLAT Gut brown soda bread	2 slices (30g per slice) or 1 thick (60g)	2
100% Spelt, sourdough	Two/three slices	minimal
Rice cake/cracker – plain/ wholewheat	4/5	2
Gluten-free bread	2 slices (30g per slice)	1
Biscuits and snacks		
Plain biscuit/shortbread	2/3 biscuits	1
Grains		
Pasta – wheat (fresh/dried)	60g dry weight	3
Pasta – gluten-free	100g dry weight	3

FRUCTAN TABLE 2: Carbohydrate foods free-from or low in FRUCTANS (can be eaten in larger quantities)
Cereals
Porridge, oat-based cereals
Oat-based muesli low in dried fruit (suitable FLAT Gut Diet recipe)
Corn and rice-based cereals (Cornflakes and Rice Krispies)
Grains
Oats – oat bran
Rice – white, basmati, brown, rice bran
Potatoes – boiled, baked, chipped, mashed, waffles, ready salted crisps
Buckwheat, millet, polenta and quinoa, and when made as pastas
Barley – sprouted
Breads
Oat bread
Breads made from rice, corn, potato and tapioca flours
100% spelt sourdough bread*
Some commercially prepared sourdoughs are not fully fermented and therefore should be counted within your fructan points. 100% sourdough spelt bread is the only option tolerated in larger quantities.
Pasta
Buckwheat pasta
Quinoa pasta
Noodles
Rice noodles, buckwheat, kelp, soba noodles
Crackers
Oat cakes
Biscuits and cakes
Oat based biscuits, flapjacks
Rice based, e.g. Rice Krispie buns or squares
Shortbread, plain biscuit, cream biscuit, chocolate biscuit, muffins, croissant, pastries, sponge and chocolate cakes may be tolerated occasionally in small quantities.

(Continued)

FRUCTAN TABLE 2: Carbohydrate foods free-from or low in FRUCTANS (can be eaten in larger quantities)
Pastry/breadcrumbs
Puff, flaky and filo pastry are normally well tolerated in small amounts
Polenta, oat, cornflake, breadcrumbs
Crumbed fish and chicken without onion or garlic are generally well tolerated. Fish in batter and tempura batter may be tolerated in small quantities.
Flour
Buckwheat, cornflour, millet, quinoa, rice, 100% spelt, sorghum, maize, potato, tapioca
Wheat flour, white, wholemeal, plain, self-raising flours may be tolerated in small quantities.
Baking ingredients
Agar-agar, arrowroot, baking powder, bicarbonate of soda, cornflour, cream of tartar

FRUCTAN TABLE 3: Carbohydrate foods high in FRUCTANS (to be avoided in Step 1 – Simplify)
Cereals
Wheat/bran-based cereals such as Weetabix, Shredded Wheat, Branflakes, All-Bran, Cheerios, wheat and fruit-based muesli, wheat germ and spelt flakes
Grains
Rye – all rye and sprouted rye
Wheat – bulgur wheat, couscous, freekeh, sprouted wheat, wheat germ
Amaranth
Barley – pearl barley
Breads
Rye bread
Rye sourdough bread*
Garlic bread
High fibre breads – very dense variety with nuts and seeds
*Rye sourdough is not tolerated.
Pasta
Spelt pasta
Gnocchi
Red lentil pasta

FRUCTAN TABLE 3: Carbohydrate foods high in FRUCTANS (to be avoided in Step 1 – Simplify)
Noodles
Processed noodle products
Crackers
Rye crispbreads
Spelt crackers
Biscuits and cakes
Fruit-filled biscuits
Wheaten biscuits
Some muesli bars
Scones
Flour
Rye, barley, coconut, soya flour

FRUCTOSE POINTS AND PORTIONS OF COMMON FOODS

FRUCTOSE TABLE

Fructose Foods Portion Control Required	Portion Size	Fructose Points: The FLAT Gut Diet Step 1 – Simplify 2–3 points daily	High Fructose Foods to avoid The FLAT Gut Diet Step 1 – Simplify
Cereals			
FLAT Gut muesli	50g	1	
Tailor-made porridge with fruit in recipe	1 portion	1	
Fresh/frozen fruit			**Fruits**
Fruits allowed in larger quantities – lemon, lime, rhubarb			**Fresh fruit** – apple, black-berries, pear, watermelon
Clementine, kiwifruit (green & gold), mandarin, satsuma, tangerine	2 average sized	1	**Stone fruits** – cherries, nectarine, peach, plum

(Continued)

FRUCTOSE TABLE

Banana, orange, starfruit	1 average sized	1		**Fresh and dried** – apricot, dates, figs, mango, goji berries, prunes, sultanas
Melon (cantaloupe, honeydew), guava, papaya (paw paw), pineapple	1 x 5cm slice (100g)	1		
Grapefruit, pomegranate, persimmon (sharonfruit)	½ average sized	1		**Agave nectar**
Blueberries, cranberries raspberries	15	1		**Honey**
Strawberries, grapes – black, red, green	10 small	1		
Lychees, kumquats, passion fruit, rambutan	4	1		
Dried fruit				
Currants, cranberries, raisins, banana chips	1 level tbsp	1		
Juice				
Orange, passion fruit	100ml	1		apple, guava, mango, pineapple, peach, pear, pomegranate, prune
Tomato	130ml	1		
Cranberry	150ml	1		
Canned fruit				
Canned fruit in water or suitable fruit juices	Amount as per fresh fruit	1		Canned fruit in apple/pear juice

LACTOSE POINTS AND PORTIONS OF COMMON FOODS

LACTOSE POINTS TABLE

Lactose Foods Portion Control Required	Portion Size	Lactose Points The FLAT Gut Diet Step 1 – Simplify 2–3 points daily	Foods low in lactose Allowed freely
Milk			**Milk**
Milk – whole, low fat, skimmed, buttermilk, goat, sheep, coconut, oat, white sauce	100ml	1	Lactose-free, almond, hazelnut, macadamia, hemp, quinoa or rice milk
Condensed, Soya* milk	40ml	1	
Milk powder – full, skimmed	15g (1 tbsp)	1	
Kefir (milk-based)	150ml	1	
Yoghurt			**Yoghurt**
Yoghurt – whole milk/low fat (natural, suitable fruit flavour)	100g	1	Lactose-free, almond, coconut, rice, soya (125g) yoghurts**
Fromage frais Greek yoghurt	130g	1	
Drinking yoghurt	180g	1	
Cheese			**Cheese**
Cheese – processed, sliced	60g (3 slices)	1	Brie, bocconcini, camembert, cheddar, edam, goat's, gouda, mozzarella, parmesan, Pecorino, soy, Swiss, manchego, gruyere, blue vein
Cheese soft/mascarpone/quark	100g	1	
Cheese – cottage***, cream cheese***	150g	1	
Ricotta***	200g	1	
Halloumi***, feta	250g	1	

(*Continued*)

LACTOSE POINTS TABLE

Cream/butter				Cream/butter
Crème fraiche – full, low fat	150ml	1		Butter
Cream – single	200ml	1		
Cream – double	250ml	1		
Cream – sour	120ml	1		
Ice cream				**Ice cream**
Ice cream – dairy	100g (2 small scoops)	1		Lactose-free, almond, coconut, oat, soya**
Chocolate				**Chocolate**
Chocolate – white/milk	50g	1		Dark chocolate
Cake				**Cake**
Cheesecake	150g	0.5		Pavlova, Eton Mess
Custard/rice pudding				**Custard/rice pudding**
Custard made with whole/ low fat milk	100ml	1		Custard/rice made with lactose-free, almond, hazelnut, macadamia, hemp, quinoa, rice milk
Rice pudding (tinned)	130g	1		

*Soya milks are tolerated in very small quantities. While low in lactose, it contains other indigestible ingredients.

**These yoghurts/ice creams may contain extra fructose or inulin, therefore may cause gut symptoms.

***For those more sensitive to lactose, these may need to be eaten in smaller quantities.

ALLIUMS TABLE		
Allium	Allowed	Avoid In Step 1
Garlic		All forms of garlic – dried, fresh, extract, powder, purée, salt and pickled
Onion		All forms – fresh, brown, Spanish, red, white, dried, extract, powder, pickled, purée and salt
Shallot		All forms
Leek	Green leaf part	White bulb part
Spring onion (scallion)	Green leaf part	White bulb part
Chives		All

FOOD FLAVOURINGS THAT ARE ALWAYS ALLOWED IN THE FLAT GUT DIET

SUITABLE FOOD FLAVOURINGS

Herbs
All fresh and dried herbs are suitable: basil, bay leaf, chive, coriander (leaves and seeds), dill, lemongrass, mint, oregano, parsley, rosemary, sage, tarragon and thyme. Sunflower, olive and rapeseed oils may be flavoured with these herbs.
Spices
Asafoetida powder, caraway seeds, cardamon, chilli, cloves, cinnamon, coriander, cumin, curry leaves, curry powder, fennel seeds, fenugreek leaves, five spice, galangal, ginger, goraka, kaffir lime leaves, mustard seeds, nutmeg, pandan leaves, paprika, poppy seeds, rampa leaves, saffron, sesame seeds, star anise and turmeric.
Flavourings
Black pepper, Bovril, capers, fish sauce, lemon juice, lime juice, miso paste, Marmite, mustard, nutritional yeasts, pepper, peanut butter, salt, soy sauce, shrimp sauce, soya sauce, spring onion (green part), tabasco, tamarind paste, vinegar (red, white, rice), vegemite, wasabi paste, Worcestershire sauce, and wheat grass.
Flavourings – suitable in small amounts
BBQ sauce, brown mustard, horseradish sauce, mayonnaise, salad cream, sweet and sour sauce, Balsamic vinegar, Dijon mustard, chutney, mint jelly, mint sauce, oyster sauce, pickles, relish, salsa sauce, tahini, yellow mustard – generally tolerated in 1–2 tablespoons.   Tomato ketchup, miso, wasabi powder – generally tolerated 1–2 teaspoons.   Coconut milk – 100ml per serving.   Many of these products vary in manufacturing, so check that they contain no garlic and no onion.
Thickening ingredients
Agar-agar, arrowroot and cornflour.
Baking ingredients
Baking powder, bicarbonate of soda and cream of tartar.
Flavoured oils
Almond, avocado, canola, coconut, olive, peanut, sesame, sunflower, rice bran, and vegetable oil, can all be flavoured with herbs and spices.   *Warning:* homemade garlic-infused oil can cause botulism. It is best to purchase commercially prepared garlic-infused oils.

PROTEIN FOODS: ANIMAL AND PLANT-BASED

Protein is an essential part of a healthy diet and is required for growth, tissue repair and to suppress appetite.

Animal proteins

All animal sources of suitable proteins:

- Beef, chicken, duck, lamb, pork (including bacon and ham – not honey roasted), turkey.
- Fish and shellfish – fresh, frozen, whitefish, oily fish, canned in brine or oil, smoked fish.
- Eggs (boiled, fried, poached, scrambled).
- Portions recommended: 75–100g of cooked meat/fish or 2 eggs per portion, twice per day.

Plant-based proteins

Plant-based proteins are becoming increasing popular with the choice of vegetarian and vegan dietary lifestyles.

- Tofu (firm is best tolerated), tempeh, quorn and spirulina are all tolerated.
- Pea protein and soy-based products may be tolerated in small amounts depending on manufacturing process, therefore eat according to personal tolerance.
- Portions recommended: 100g per portion, twice per day.
- Protein-rich beans and legumes are poorly digested and only suitable in limited amounts. Tinned legumes are better tolerated than soaked and boiled legumes. Further details are available in the table in the vegetables section.

Nut and seed proteins

Nut and seed proteins are suitable in a small handful. Nut butters are also well tolerated. There are only two nuts / nut butters that should be avoided in step one - Simplify, which are cashew and pistachio as these are high in fermentable components (GOS).

SUITABLE BEVERAGES FOR THE FLAT GUT DIET

Beverages
The FLAT Gut Diet Step 1 – Simplify

Allowed	Moderation	Avoid
Water – natural Flavoured water – low fructose fruits	Coconut water	Smoothies with unsuitable vegetables (see table)
		Fruit juices high in fructose
Squashes – reduced sugar, no added sugar – lemon, lime, orange, pineapple, tropical, strawberry	Isotonic sports drinks – (6–8g carbohydrate per 100ml) Lucozade Sport, Gatorade, Powerade, Energise Sport	Hypertonic sports drinks – (greater than 10g carbohydrate per 100ml) such as Lucozade Energy
Lemonade, club orange, 7Up, ginger beer, sparkling water, soda water*, tonic water*		Sports drinks – containing fructose
Coffee** – decaffeinated, espresso, instant, regular, white***. Almond, hazelnut, lactose-free and rice milk lattes are tolerated	Powder drinks – drinking chocolate, cocoa powder, malted chocolate-flavoured – 2 heaped teaspoons carob powder 1 tsp	Avoid regular milk and skimmed milk latte (milk from lactose points)
Tea** – black, decaffeinated, green, rooibos, white**		
Herbal teas – chai/chamomile (weak), licorice, dandelion, lemon, ginger, peppermint and tumeric		Herbal and fruit teas – apple, pear, blackberry, chicory, fennel, oolong
Carbonated beverages*		
Dioralyte – electrolyte drinks		

*Limit intake of carbonated drinks.

**Caffeine may trigger diarrhoea.

***Milk and latte from lactose daily points.

SUITABLE ALCOHOLIC DRINKS FOR THE FLAT GUT DIET

ALCOHOL

The FLAT Gut Diet Step 1	The FLAT Gut Diet Step 2	The FLAT Gut Diet Avoid in all steps
Minimize your alcohol intake to observe true IBS/functional dyspepsia symptoms	Small amounts of alcohol* Beer Gin Vodka Wine, red Wine, sparkling Wine, white Whiskey	Cocktails with fruit juices Cider Rum Wine, dessert sweet

*please see our advice about safe drinking of alcohol in Chapter 13.

SUITABLE SWEETENERS, PRESERVES AND SPREADS

Sweeteners, Preserves and Spreads

The FLAT Gut Diet	The FLAT Gut Diet
Allowed	Avoid
Glucose – regular, syrup	Honey
Dextrose	Agave – syrup, nectar
Sugar (sucrose) – brown, palm, white	Fructose – syrup, corn syrup, corn syrup solids, fructose-glucose, glucose-fructose syrup, fruit juice concentrate
Preserves – jam, marmalade	Sugar-free – jams, marmalades, chewing gum, mints, sweets, chocolates
Spreads – milk chocolate spread, hazelnut and vanilla, Marmite, Nutella, peanut butter	Polyol sweeteners (ending in 'ol'), Erythritol, sorbitol, mannitol, maltitol, xylitol or isomalt
Syrup – golden, maple, rice, malt	Fructan ingredients can be found in some foods such as protein bars, cereal bars, low calorie snack bars and yoghurts. These foods will be listed as ingredients such as Fructo-oligosaccharides (FOS), oligofructose, inulin (chicory root)

Sweeteners, Preserves and Spreads

The FLAT Gut Diet	The FLAT Gut Diet
Treacle	
Artificial sweeteners – aspartame, saccharin, sucralose, Canderel, Splenda, Hermesetas, Stevia	

21

The FLAT Gut Diet: recipes, snacks and beverages

In the following pages you will find the recipes for the main meals of your 7-Day diet plan for Step 1 of the FLAT Gut Diet. We have designed these recipes to be as tasty as possible while also helping to minimize your gut symptoms. We have also included breakfast recipes that require a little advance preparation.

In addition, you will find the recipes for some easy-to-prepare beverages and snacks that will not trigger your IBS symptoms. There are also some additional taster recipes.

Sample meal plan recipes

BREAKFAST

Tailor-made porridge

Porridge is a brilliant way to start the day, particularly in the colder months of the year. This recipe uses milk but you can also make it with water if you prefer. It is important that the cooked porridge has a nice soft dropping consistency, so add a little more water if you think it needs it. Use the table below to mix and match the basic recipe and add your favourite toppings.

Serves 1

INGREDIENTS

40g porridge oats

100ml whole/low-fat milk (from lactose points) or use 150ml of lactose-free milk or 150ml of almond, hazelnut, macadamia, rice or hemp milk for a dairy free alternative

15 blueberries/raspberries or 10 small strawberries

2 teaspoons maple syrup

2 teaspoons toasted chopped hazelnuts

METHOD

Place the porridge and milk in a heavy-based pan with 60ml of water over a medium to high heat and bring to the boil. Then reduce the heat to low. Add a pinch of salt and cook for eight to ten minutes until you have a thick and creamy texture, adding a little more water if you think it needs it.

Pour the porridge into a bowl and scatter over the fruit of your choice. Drizzle with the maple syrup and scatter over the hazelnuts to serve.

Breakfast recipe tip: oats

Oats are not tolerated by everyone and can be problematic for some people. You will need to find your tolerance level: one portion of oats per day in cereal, bars, breads or oat cakes is normally well tolerated. Also, some people may experience gut symptoms with successive intake of oats over several days. You can always substitute oats with rice, quinoa or buckwheat flakes.

Fruit, spice and nut options for porridge and muesli

Variety is the spice of life and good for gut microbial diversity. Choose from this selection of fruit, nuts and spices when preparing your porridge or muesli.

Foods	Nuts (whole/ chopped)	Seeds*	Spices	Butters	Fruits: choose one of the following
Porridge	Almonds 10	Chia	Cinnamon	Almond	Blueberry, cranberry/ raspberry 15
Muesli	Brazil	Poppy	Cloves	Peanut	Grapes 10 small

(*Continued*)

Foods	Nuts (whole/ chopped)	Seeds*	Spices	Butters	Fruits: choose one of the following
	Hazelnuts 10	Pumpkin	Ginger		Kiwi 2
	Macadamia	Sesame	Nutmeg		Pomegranate ½
	Peanuts	Sunflower	Turmeric		Strawberries 10 small
	Pecan	Linseed Flaxseed			Dried currants, cranberries, raisins 1 level tbsp
	Walnuts				

*Seeds can increase fibre intake significantly. 1 tablespoon chia seeds is 6 fibre points, 1 tablespoon linseed/flaxseed is 4 fibre points, 1 tablespoon of almond butter is 3 fibre points and 1 tablespoon of peanut butter is 1 fibre point. If you have IBS–D, you may not tolerate the extra fibre.

FLAT Gut muesli

This is a great recipe to batch prepare so that you have a delicious and nutritious breakfast that needs no preparation. Muesli is easy to make so don't be intimidated by the length of the ingredients list. The mixture will fill a 1.2 litre Kilner jar and is enough to keep you going for a couple of weeks. If you prefer, make it up with quinoa pops or buckwheat or rice flakes, or you could even experiment with a mixture.

Serves 20

INGREDIENTS

500g porridge oats

½ teaspoon ground cinnamon

1 teaspoon vanilla extract

50g flaked almonds

50g sunflower seeds

50g pumpkin seeds

50g skinned hazelnuts, chopped

50g dried cranberries

75g toasted shredded coconut

50g dried banana chips

50g raisins

25g linseed (optional)

sliced banana, halved strawberries and/or seedless grapes, to garnish

100ml whole or low-fat milk (from lactose points), if not enough top up with lactose-free milk) or lactose-free milk or hazelnut, almond, rice or hemp milk for a dairy-free alternative or 100g of natural yoghurt or 125g coconut yoghurt, to serve

METHOD

Preheat the oven to 200C/400°F. Put the oats into a bowl and sprinkle over the cinnamon and vanilla extract, tossing to coat evenly, then spread them out on a baking sheet.

On a separate baking sheet, sprinkle over the flaked almonds, sunflower seeds, pumpkin seeds and hazelnuts, and place on the top shelf of the oven with the porridge oats on the shelf underneath. Bake for ten to 15 minutes until lightly toasted, tossing occasionally so that they cook evenly. Remove both baking sheets from the oven and leave to cool.

Once cool, place the toasted porridge oats, nuts and seeds in a large bowl. Stir in the cranberries, coconut, banana chips, raisins and linseed, if using. Transfer to a 1.2 litre Kilner jar or airtight container and seal tightly with a lid. It will keep for up to two weeks.

To serve, weigh a 50g portion of the muesli into a bowl and scatter the sliced banana, strawberries and/or grapes on top to garnish. Pour over the milk or add a dollop of yoghurt.

FLAT Gut breakfast bar

These bars are full of goodness and guaranteed to get you off to a good start.

Makes 8 bars

INGREDIENTS

100g porridge oats

70g plain flour*

½ teaspoon ground cinnamon

good pinch ground nutmeg

¼ teaspoon salt

¼ teaspoon bicarbonate of soda

120ml coconut oil

5 tablespoons maple syrup

1 teaspoon vanilla extract

1 large egg

50g unsweetened desiccated coconut

100g toasted chopped hazelnuts

100g pumpkin seeds

100g sunflower seeds

*Option: 100% sieved spelt flour can be used as an alternative.

METHOD

Preheat oven to 180C/350°F. Line a 20cm baking tin with parchment paper. Place the porridge oats, flour and spices in a bowl with the salt and bicarbonate of soda, stirring to combine.

Place the coconut oil in a Pyrex jug and microwave for 30 seconds until it melts (or use a small pan over a low heat). Add the maple syrup and vanilla. Break in the egg and whisk to combine. Stir the mixture into the dry ingredients and then fold in the desiccated coconut, hazelnuts, pumpkin and sesame seeds.

Press the dough into the baking tin, wetting your hands a little with water if needs be to help pat the dough down evenly. Bake for 15 minutes. Cool bars in tin, then slice and store in an airtight container for up to three days. Use as required.

FLAT Gut oat shake

This shake will take no more than two minutes to prepare. The shake is tasty, refreshing and easy to make. Use fresh or frozen soft fruits, plus maple syrup for a little sweetness and oats for slow-release fuel. Government guidelines recommend that you boil frozen berries for one minute before using but this does not apply if you purchase fresh berries and freeze them for a later date.

Serves 1

INGREDIENTS

45g porridge oats

15 fresh or frozen raspberries/blueberries or 10 small strawberries

100ml whole or low-fat milk or use 150ml of lactose-free milk or 200ml of hazelnut, almond, rice or hemp milk for a dairy-free alternative

2 teaspoons maple syrup

1 teaspoon vanilla extract

1 teaspoon chia seeds

100g ice cubes

METHOD

Place all the ingredients in a blender or Nutri-bullet with 100ml of water. Pour into a large glass or beaker to serve.

Blueberry & banana power pancakes

Pancakes have to be a Sunday morning favourite. Of course, you can now easily buy oat flour instead of making your own, but it is so easy and means you do not have to buy a special ingredient for this treat!

If you want to bump up the goodness add a teaspoon of chia seeds, flax or linseeds.

Serves 4

INGREDIENTS

2 small firm bananas

4 large eggs, separated

4 tablespoons regular or low-fat milk (or use your favourite lactose-free milk)

120g porridge oats

¼ teaspoon ground cinnamon

1 teaspoon baking powder

2 tablespoons sunflower oil

30 blueberries

2 teaspoon toasted chopped hazelnuts

4 tablespoons maple syrup

METHOD

Peel and mash one of the bananas and mix in a bowl with the egg yolks until evenly combined, then stir in the milk.

Blitz the porridge oats until fine in a food processor or blender, and then stir into the banana mixture with the cinnamon and baking powder.

In a separate bowl, whisk the egg whites until you have achieved soft peaks and fold gently into the batter.

Heat a non-stick frying pan over a medium heat and add some of the oil. Add three spoonfuls of the mixture and cook for about two minutes on each side until golden brown. Flip over and cook for another minute. Keep the pancakes warm while you cook the remainder – you'll make 12 small pancakes in total.

Meanwhile, peel and thinly slice the remaining banana. Layer up the pancakes on plates with the bananas and blueberries and scatter over the hazelnuts, then drizzle with the maple syrup to serve.

LIGHT MEAL

Poached eggs with roasted vine tomatoes and smashed avocado

This has become a very trendy breakfast that you'll probably pay about a tenner for in a nice café, but it really is so easy to make at home. Don't be tempted to add salt to the water before poaching the eggs as it toughens the white.

Serves 4

INGREDIENTS

4 vines of cherry tomatoes

a little extra-virgin olive oil

½ teaspoon fresh thyme leaves

1 tablespoon white wine vinegar

4 large eggs

8 slices of wholemeal/FLAT Gut brown soda bread or 4 slices of 100% spelt sourdough bread

1 firm ripe avocado

½ lemon

sea salt and freshly ground black pepper

METHOD

Preheat the oven to 200C/400°F. Arrange the tomatoes in a baking tin and drizzle over a little oil, then season with salt and pepper and sprinkle with thyme. Roast for 10–15 minutes until slightly charred but still holding their shape.

Fill a large pan with water, add the vinegar and bring to a hard boil over a high heat. Crack in the eggs and cook for two and a half minutes, then drain on kitchen paper.

Toast the bread in a toaster or on the griddle and drizzle with a little olive oil, then arrange on warmed plates. Mash the avocadoes in a bowl with a squeeze of lemon and season to taste and then divide among the toast and put a poached egg on each one. Add the roasted tomatoes to serve.

Spiced carrot & ginger soup

This is a handy and healthy soup that is easy to make and actually does not need any stock, which is a major bonus. It also has the most fantastic vibrant colour and costs little or nothing to make. What more could you want?

Serves 4–6

INGREDIENTS

knob of butter

2 celery sticks, thinly sliced

20g root ginger, peeled and finely chopped

675g carrots, grated

2 teaspoons ground cumin

1 tablespoon maple syrup

1 teaspoon fresh lemon juice

900ml boiling water

1 tablespoon soured cream

1 tablespoon chopped fresh coriander (optional)

sea salt and freshly ground white pepper

METHOD

Melt the butter in a large pan. Add the celery and ginger, then cook gently for four to five minutes until softened but not coloured, stirring occasionally. Stir in the carrot, cumin, maple syrup and lemon juice and season with a pinch of salt and plenty of pepper. Pour the boiling water into the carrot mixture and bring to the boil. Reduce the heat and simmer for 45 minutes until slightly reduced and the carrots are tender.

Remove the soup from the heat and blitz with a hand blender until smooth. To serve, ladle into warmed bowls. Drizzle a little soured cream into each one and garnish down the middle with a thin line of coriander, if liked.

If you want to make up batches of soup and freeze down portions for convenience, simply pour the soup into a jug and use to fill a Zip-Lock or suitable freezer bags. Let out the air and lay flat to freeze so that they don't take up too much room in the freezer.

FLAT Gut brown soda bread

This is another great bread for novice cooks to start with as there is no proving or kneading, as the bicarbonate is responsible for the rise when it reacts with the buttermilk. If you don't have any buttermilk in the house, use sour full-fat milk with the juice of half a lemon or one tablespoon of red wine vinegar, which gives the bread a little more tang. The addition of the seeds is a personal preference so feel free to leave them out if you want a more traditional brown soda.

Makes 1 x 600g loaf

INGREDIENTS

olive oil, for greasing

225g white flour, plus extra for dusting*

225g wholemeal flour*

1 teaspoon bicarbonate of soda

1 teaspoon fine salt

400ml buttermilk, plus extra if needed

1 tablespoon melted butter

1 tablespoon maple/golden syrup

handful of porridge oats

1 dessertspoon of sunflower, pumpkin or sesame seeds (optional)

*this can be replaced with 100% sieved spelt white and wholemeal flour.

METHOD

Preheat the oven to 200C/400°F. Lightly oil a 600g non-stick loaf tin. Sift the flours into a large bowl, then stir in the bicarbonate of soda and salt. Make a well in the centre and add the buttermilk, melted butter, seeds (if using) and golden syrup. Using a large spoon, mix gently and quickly until you have achieved a nice dropping consistency. Add a little bit more buttermilk if necessary, until the dough binds together without being sloppy.

Put the mixture into the prepared loaf tin and sprinkle over the porridge oats with the sunflower, pumpkin and sesame seeds, if using. Bake for 40 minutes, until cooked through and the loaf has a slightly cracked, crusty top, checking halfway through the cooking time to make sure that it isn't browning too much. If it is, reduce the temperature or move the loaf down in the oven.

To check that the loaf is properly cooked, tip it out of the tin and tap the base – it should sound hollow. If it doesn't, return it to the oven for another five minutes. Tip out onto a wire rack and leave to cool completely, then cut into slices to serve. The brown soda bread is best used on the day it is made but it will keep for two to three days.

Tailor-made sandwich

Choose topping/filling

Tomato, mozzarella and basil	Tuna and mayonnaise
Ham, cheese and tomato	Egg mayonnaise
Cheese and pickles	Feta, olives, tomato, basil
Chicken and bacon	Rasher (bacon or turkey) and tomato
BLT	Salad, lettuce, tomato, cucumber, grated carrot, peppers, radish, beetroot (small amount)
Salmon, mayonnaise and dill	Sardines, mackerel
Prawn cocktail	Crab, crème fraiche with chopped chives and cherry tomatoes

Bread tips

Rye sourdough is not tolerated. Some commercially prepared sourdoughs are not fully fermented, therefore should be counted within your fructan points. One hundred per cent sourdough spelt bread is the only option tolerated in larger amounts. Spelt bread with honey is not well tolerated.

Breads made with sieved 100% spelt white or wholemeal flour are better tolerated and, as mentioned, can be substituted in recipes.

Mediterranean Greek-style salad with baked feta

This is a twist on the traditional Greek salad that is perfect to serve in the middle of the table for a relaxed lunch or as a starter for a smart dinner. The baked feta adds a lovely contrast to the warm and salty cheese infused with chilli and oregano.

Serves 2–4

INGREDIENTS

200g block feta cheese

1 mild red chilli, seeded and finely chopped

½ teaspoon fresh oregano or ½ teaspoon dried

1 tablespoon olive oil

1 cucumber

250g mixed coloured cherry tomatoes

100g pitted black Kalamata olives

1 teaspoon red wine vinegar

1 tablespoon extra-virgin olive oil

sea salt and freshly ground black pepper

METHOD

Preheat the oven to 200C/400°F. Place the feta in a small baking tin that has been lined with parchment paper and sprinkle over the chilli and oregano. Drizzle with the olive oil and season with pepper. Bake for ten minutes until heated through.

Meanwhile, cut the cucumber in half lengthways and using a teaspoon scoop out the seeds, then thinly slice on the diagonal. Cut the cherry tomatoes in half and place in a bowl. Add the olives and drizzle over the vinegar and extra-virgin olive oil. Season with a pinch of salt and plenty of pepper. Then toss lightly to combine and arrange on a platter.

When the feta cheese is ready, place at one end of a platter and spoon the salad at the other end to serve.

Tuna mayonnaise and gherkin salad

Super simple, grab-and-go lunch rich in omega 3s.
Serves 2

INGREDIENTS

2 x 145g tins of tuna, drained

2 tablespoon mayonnaise

2 gherkins, finely chopped

½ lemon, juiced

2 baby gem lettuce

Tomatoes, cucumber or grated carrot, sliced radish

METHOD

Mix together the tuna, mayonnaise, gherkins, lemon juice and plenty of black pepper, then place on lettuce leaves. Serve with the FLAT Gut Diet brown soda bread and with your chosen low fermentable salad.

Prawns and red cabbage on sourdough

Prawns and slaw on sourdough is a gut-friendly winner. Sourdough 100% spelt is one of the easiest breads to digest, made using a slow process of fermentation.
Serves 2

INGREDIENTS

100g red cabbage, trimmed and very finely sliced

1 small carrot, scrubbed and coarsely grated

1 tablespoon mixed seeds

2 slices of 100% spelt sourdough bread or 1 thick or 2 thin slices of FLAT Gut brown soda bread (60g total) or 2 slices of wholemeal/ wholegrain bread

A handful of watercress or baby salad leaves

75g cooked and peeled prawns, thawed if frozen and drained

Freshly ground black pepper

FOR THE DRESSING:

2 tablespoon extra-virgin olive oil

1 teaspoon fresh lemon juice, plus extra to serve

½ teaspoon Dijon mustard

½ teaspoon maple/golden syrup (optional)

METHOD

To make the dressing, whisk together the olive oil, lemon juice, mustard and maple/golden syrup (if using) in a bowl. Add the cabbage, carrot and mixed seeds to the bowl and toss together.

Divide the bread between 2 plates and top with the watercress or salad leaves.

Place some of the slaw on top, then scatter with the prawns. Squeeze a little extra lemon juice over the prawns and season with black pepper to serve.

Oriental layered salad in a jar

The crunch and texture of a salad can be lost in transportation or if you want to make it in advance, so why not assemble it in individual jars and make the dressing separately. Many supermarkets now sell spiralized vegetables which can save you time, but of course you can also prepare your own. If you don't have a spiralizer then just grate the vegetables or cut into julienne.

Serves 1

INGREDIENTS

50g spiralized carrots

50g cooked rice noodles

50g red cabbage, tough core removed and finely shredded

25g spiralized courgettes

1 small yellow pepper, seeded and diced

25g fresh beansprouts

1 baby Hass avocado (or use ¼ of regular sized one – 45g)

small handful baby spinach leaves

1 tablespoon soy sauce

3 tablespoons rice vinegar

1 teaspoon toasted sesame oil

METHOD

Take a jar that is about 600ml in size and put the carrot in the bottom. Add a layer of the noodles. Cover with the red cabbage and then add a layer of the courgettes. Sprinkle the yellow pepper on top followed by the beansprouts.

Cut the baby avocado in half and remove the stone, then peel off the skin and dice the flesh. Add to the salad and finish with the spinach.

Make the salad dressing with the soy sauce, rice vinegar and sesame oil. When ready to serve use to dress the salad.

Scrambled eggs with smoked salmon

If you prefer your scrambled eggs chunkier, don't whisk the egg and cream mixture – pour or break the eggs straight into the pan and then add the cream, stirring continuously. Leave it to set for a minute or so and then give it a gentle stir, barely breaking up the egg mixture so it stays very fluffy and light.

Serves 4

INGREDIENTS

4 eggs

2 tablespoons cream

1 tablespoon snipped fresh chives

1 tablespoon olive oil

300g smoked salmon ribbons

4–8 slices of the FLAT Gut soda bread (1 thick slice or 2 thin slices per person – 60g total)

sea salt and freshly ground black pepper

fresh long chives, to garnish (optional)

METHOD

Preheat the grill. Whisk together the eggs, cream, chives and plenty of freshly ground black pepper. Heat a drizzle of olive oil in a non-stick frying pan. Add the egg mixture and whisk continuously for two to three minutes, until just set but still soft. Remove from the heat, as it will continue to cook. Check the seasoning and add a pinch of salt if you think it needs it.

Meanwhile, toast the bread on a griddle pan and then drizzle each piece of toast with the rest of the olive oil and cut into pieces.

Arrange the griddled bread on plates and top each one with the scrambled eggs and arrange the smoked salmon alongside. Garnish with the chives, if using, to serve.

MAIN MEAL

Chinese-style chicken with Asian greens

This stir-fry is incredibly easy to master once you follow some basic rules. Invest in a wok if you don't have one and have everything ready before you start to cook.

Serves 4

INGREDIENTS

200g wholegrain rice

600ml water

400g chicken stir-fry strips

1 tablespoon soy sauce

2 teaspoons Shaoxing rice wine or dry sherry

1 teaspoon sea salt

1 teaspoon freshly ground black pepper

1 tablespoon toasted sesame oil

2 teaspoons cornflour

450g selection prepared Asian green vegetables, such as baby pak choi, baby bok choy and choy sum (cut into 4cm pieces on the diagonal)

1 tablespoon olive oil

2 tablespoons finely chopped spring onion (green part only)

2 tablespoons finely shredded fresh root ginger

2 red Thai chillies, seeded and sliced

2 tablespoons Thai fish sauce (nampla)

3–5 tablespoons chicken stock (IBS-friendly, see 'Stocks', below)

2 tablespoons oyster sauce

METHOD

Rinse the rice well in a sieve and then put into a pan with the water and a pinch of salt. Bring to the boil, then reduce the heat and cover with a lid. Simmer for 25 minutes without lifting off the lid, then turn off the heat and leave to sit for another ten minutes for perfectly cooked rice (or simply cook according to packet instructions).

Place the chicken in a non-metallic bowl and add the soy sauce, rice wine or sherry, salt, half a teaspoon of the pepper, one teaspoon of the sesame oil and the cornflour. Mix well to combine and set aside for 30 minutes to allow the flavours to develop.

Heat a wok over a high heat until it is hot. Add the oil, and when it is slightly smoking, add the spring onions, ginger and the remaining half a teaspoon of pepper. Stir-fry for a few seconds, add the chicken, then stir-fry for two to three minutes until lightly browned.

Add the prepared vegetables, chillies and fish sauce to the pan, adding the stock as needed. Stir-fry on a medium to high heat for three to four minutes, or until the chicken is cooked and the vegetables are tender. Add the oyster sauce and the remaining sesame oil and stir-fry for another minute. Divide the rice among warmed bowls and top with the stir-fry to serve.

Sesame-crusted salmon with pickled vegetable salad

Salmon is a great source of omega 3s and high in protein to fill you up. If you are really short of time, pick up a couple of packets of carrot spirals for ease. Alternatively, you can make your own with a spiralizer, which is a nifty little gadget that is very inexpensive to buy.

Serves 4

INGREDIENTS

5 tablespoons rice wine vinegar

5 tablespoons water

1 tablespoon caster sugar

4 carrots, peeled and julienned (or use a shop-bought packet of carrot spirals)

1 mild red chilli, seeded and finely chopped

1 heaped tablespoon white sesame seeds

1 heaped tablespoon black sesame seeds

1 tablespoon olive oil

4 x 150g boneless salmon fillets, skin on but scaled

1 x 200g packet of baby cucumbers

2 little gem lettuces, shredded

2 spring onions, finely chopped (green parts only)

1 tablespoon chopped fresh coriander

sea salt and freshly ground black pepper

500g baby new potatoes

METHOD

Place the potatoes in a pan of cold water over a high heat and add a pinch of salt. Bring to the boil and cook for 12–15 minutes until tender. Drain and set aside.

Put the rice wine vinegar in a pan with the water and sugar. Add a pinch of salt and bring to a simmer. Put the carrots and chilli in a bowl, then pour the vinegar on top. Set aside to pickle.

Meanwhile, place the black and white sesame seeds on a plate and mix well to combine. Press in each salmon fillet, skin side down, until evenly coated in the sesame seed mixture.

Heat a large non-stick frying pan over a medium heat. Add the olive oil and swirl it around and then add the salmon fillets and cook, skin side down, for three minutes. Turn the salmon fillets over and cook for another two minutes, or until cooked to your liking.

Pare the cucumbers into ribbons and put into a bowl with the lettuce, spring onions and coriander. Drain off the pickled carrot and chilli and add to the bowl along with a good grinding of pepper, then toss to lightly coat.

Arrange the pickled vegetable salad on plates and top each one with a piece of sesame-crusted salmon. Serve with baby boiled potatoes, with an optional knob of butter and freshly chopped parsley or chives.

Sweet & sour tofu (or poultry) & vegetable stir-fry

The crispy tofu and crunchy vegetables are an excellent combination in this stir-fry. Make sure you have everything prepped before you start cooking.

Serves 4–6

INGREDIENTS

350g medium rice noodles (Pad Thai)

40g macadamia nuts

4 spring onions, finely chopped (green part only)

2 small carrots, thinly sliced

1 small courgette, trimmed and thinly sliced

200g tender-stem broccoli, sliced on the diagonal into small pieces

2 tablespoons olive oil

400g firm tofu, drained and cut into bite-sized pieces (or 400g of poultry)

2 tablespoons maple syrup

3 tablespoons soy sauce

2 tablespoons rice vinegar

300g tin pineapple chunks in natural juice, drained

sea salt and freshly ground black pepper

METHOD

Put the noodles in a large flat dish and cover with boiling water. Follow the rice noodle packet cooking instructions, then drain and rinse under cold running water and leave in cold water to prevent them from sticking.

Heat a wok or non-stick frying pan over a medium heat. Tip in the macadamia nuts and dry cook until evenly toasted, tossing them regularly to ensure they do not catch and burn. Tip on to a chopping board and leave to cool, then roughly chop. Set aside until needed.

Reheat the wok over a medium-high heat. Add half the oil, swirling up the sides. Using tongs, add the tofu. Season with a pinch of salt and plenty of pepper and then stir-fry for three to four minutes until cooked through and lightly browned. Transfer to a plate.

Wipe out the wok and reheat it over a medium heat. Add the rest of the oil, coating up the sides. Tip in the spring onions and stir-fry for

30 seconds until sizzling. Add the carrots, courgette and tender-stem broccoli and stir-fry for another two to three minutes until almost tender but still with some crunch.

Add the tofu (or chicken) and pineapple to the vegetable mixture and toss until well combined, then drizzle over the honey, soy and vinegar and cook for one to two minutes, tossing until everything is evenly coated.

Divide the noodles between warmed bowls and spoon the sweet and sour vegetables on top. Scatter over the macadamia nuts to serve.

Keralan chicken curry with green beans

This curry originates from the tropical south coast of India and is traditionally served with jasmine rice and green beans. It is wonderfully light and fragrant.

Serves 4–6

INGREDIENTS

350g Thai jasmine rice

600ml boiling water

1 tablespoon olive oil

2 spring onions, finely minced (green part only)

2.5cm piece fresh root ginger, peeled and finely grated

1 cinnamon stick

6 whole cardamom pods, lightly crushed to split open

½ teaspoon ground turmeric

1 teaspoon chilli powder

20g fresh coriander

200ml tin coconut milk

120ml chicken or vegetable stock (IBS-friendly, see 'Stocks', below)

pinch of sea salt

400g boneless chicken, cut into wafer-thin slices

80g mangetout

80g sugar snap peas

150g extra fine green beans

1 lime, cut into wedges

METHOD

To prepare the jasmine rice, place it in a heavy-based pan with the boiling water. Bring back to the boil and then cover with a tight-fitting lid and reduce the heat to low. Simmer for 15–20 minutes, or until all of the water has been absorbed and the rice is tender.

Meanwhile, heat the oil in a heavy-based pan over a medium heat. Add the spring onions, ginger, cinnamon and cardamom pods and stir-fry for 20 seconds. Stir in the turmeric and chilli powder and cook for 30 seconds to one minute until fragrant, stirring.

Reserve a few coriander sprigs for garnish and blend the remainder with the coconut milk in a mini food processor or with a hand blender. Stir into the pan with the stock and bring to the boil. Reduce the heat and simmer for one to two minutes, then stir in the chicken and vegetables and season with the salt. Very thinly sliced chicken should take just minutes to cook; typically three to four minutes. Continue to simmer for two to three minutes (cook for longer if using thicker pieces of chicken) until the chicken and vegetables are tender, stirring constantly. Add the lime juice to taste.

Remove the lid from the rice and fluff up the grains with a fork, then divide among warmed bowls and ladle over the Keralan chicken curry. Garnish with the reserved coriander sprigs and lime wedges to serve.

Spaghetti Bolognese

This is a brilliant ragu mince recipe, which can be made in large quantities so that you can freeze it in individual batches to use when the need arises. Feel free to bulk it out with extra vegetables such as a couple of courgettes or peppers, adding them along with the carrots and celery at the beginning of the recipe. Alternatively serve it with courgette spaghetti, which is very easily made using a spiralizer and is also stocked ready-made by many supermarkets.

Serves 4

INGREDIENTS

2 tablespoons olive oil

150g rindless smoked streaky bacon, diced

1 celery stick, finely diced

2 carrots, finely diced

½ teaspoon fresh thyme leaves or ½ teaspoon dried

1 bay leaf

½ teaspoon dried oregano

400g lean minced beef

1 tablespoon tomato purée

150ml dry white wine

400g tin Italian chopped tomatoes

1 teaspoon Worcestershire sauce

600ml chicken stock (IBS-friendly, see 'Stocks', below)

240g fresh egg pasta dry weight or 300g buckwheat pasta dry weight

2 tablespoons freshly grated Parmesan

sea salt and freshly ground black pepper

METHOD

Heat a large, heavy-based pan over a medium heat. Add the olive oil and tip in the bacon. Cook for a couple of minutes until the bacon is crispy and has released some natural fats, then add the celery, carrots, thyme, bay leaf and oregano and cook over a medium heat, stirring occasionally, until the vegetables have softened and taken on a little colour.

Add the minced beef to the pan. Mix to combine, then sauté until well browned, breaking up any lumps with a wooden spoon. Stir in the tomato purée and cook, stirring for a minute or two.

Deglaze with a little wine, scraping up any sediment. Pour in the remaining wine with the tomatoes, Worcestershire sauce and stock and season with a pinch of salt and plenty of pepper. Bring to the boil, then reduce the heat and simmer, stirring from time to time, for one hour (or up to four hours is fine) until the beef is meltingly tender – the longer you cook it the more tender it becomes.

When ready to serve, bring a large pan of water to a rolling boil. Add a good pinch of salt and swirl in the spaghetti. Stir once, then cook for eight to ten minutes, or according to the instructions on the packet, until the pasta is al dente (tender but still firm to the bite). Drain and return to the pan with the Bolognese sauce. Toss until well combined. Divide the spaghetti Bolognese among warmed wide-rimmed bowls and sprinkle over the Parmesan to serve.

Roasted aubergine with cherry tomatoes & goat's cheese

It's not without reason that in the Middle East and throughout Italy aubergines are still regarded as the poor man's meat. Prepare this in advance and simply pop it in the oven just before you are ready to serve.

Serves 4

INGREDIENTS

2 large aubergines

3 tablespoons olive oil

1 red pepper

2 spring onions, finely chopped (green parts only)

small handful fresh basil leaves, torn

225g cherry tomatoes, halved

2 x 100g individual goat's cheeses, thinly sliced (with rind)

sea salt and freshly ground black pepper

tossed green salad, (optional add some toasted seeds) to serve

METHOD

Preheat the oven to 200C/400°F. Cut the aubergines in half and trim off the stalks. Brush the cut sides with a little of the oil and season with a pinch of salt and plenty of pepper, then place in a roasting tin with the red pepper and bake for 30–35 minutes or until the flesh of the aubergine is tender and the skin of the red pepper is blackened and blistered.

Meanwhile, heat a tablespoon of the oil in a non-stick frying pan set over a medium heat and sauté the spring onions for 30 seconds to one minute until softened but not browned. Set aside.

Remove the roasted vegetables from the oven (leaving the oven turned on), and once the red pepper is cool, peel away the skin and finely chop the flesh, discarding the seeds. Place in a bowl with the sautéed spring onions. Scoop out the flesh from the aubergines, to within 1cm of the skin and finely chop the removed flesh. Add to the red pepper and spring onions, then add the basil and season with a pinch of salt and plenty of pepper.

Pile the mixture back into the aubergine shells and arrange the cherry tomatoes and slices of goat's cheese on top. Drizzle over the remaining oil and return to the oven for 20–25 minutes until the cherry tomatoes are lightly charred and the goat's cheese is bubbling. Arrange on plates with the salad to serve.

Sirloin steak with Cashel blue butter & roasted root vegetables

It is a good idea to bring steaks to room temperature by removing them from the fridge 30 minutes before you want to cook them. This flavoured butter is a serious treat so feel free to leave it out if you are watching what you eat.

Serves 4

INGREDIENTS

25g butter (at room temperature)

50g Cashel blue cheese (at room temperature)

1 tablespoon chopped fresh flat-leaf parsley

2 tablespoons olive oil, plus extra for brushing

2 potatoes cut into chunks

1 small swede, cut into chunks

2 carrots, cut into chunks

2 small parsnips, cut into chunks

1 leek, thinly sliced (green part only)

4 x 175g sirloin steaks

sea salt and freshly ground black pepper

METHOD

Preheat the oven to 180C/350°F. To make the Cashel blue butter, place the butter in a bowl with the Cashel blue cheese and parsley. Season with pepper and mash with a fork until well mixed. Spoon on to a small square of parchment paper and shape into a cylinder, about 2.5cm and twist the ends. Chill for at least one hour to firm up.

Place the two tablespoons of olive oil in a roasting tin with the potatoes, swede, carrots and parsnips and toss well to combine. Season with a pinch of salt and plenty of pepper and roast for 30 minutes or until the vegetables are almost but not quite tender, tossing occasionally. Add the leek to the roasting vegetable mixture, tossing to ensure that they are evenly combined. Roast for another 15–20 minutes or until all of the vegetables are completely tender and just beginning to caramelize.

Meanwhile, heat the large griddle or non-stick frying pan over a high heat. Brush the steaks lightly with oil and then season with pepper. Add to the heated pan and cook for two minutes on each side, then reduce the heat and cook the steaks for five to ten minutes, turning once, depending on how rare you like them. Transfer the steaks to warmed plates and season with salt. Unwrap the Cashel blue butter and cut into four pieces, then place one on each steak and set aside in a warm place to rest. Add the roasted root vegetables to serve.

DESSERT

Eton Mess with dark chocolate shavings

This traditional English dessert is a mixture of strawberries, meringue and cream. It literally takes minutes to prepare and needs to be served immediately or the crunch from the meringue will be lost.

Serves 4

INGREDIENTS

320g strawberries, hulled

400ml cream

4 meringue nests (shop-bought or homemade)

50g piece dark chocolate (at least 70% cocoa solids)

METHOD

Place half the strawberries in a blender and blend to a puree. Chop the remaining strawberries into small pieces.

Pour the cream into a bowl and whip until stiff peaks form. Break the meringue into bite-sized pieces and gently fold into the cream with the strawberry puree so that you can still see visible swirls. Finally, fold in the chopped strawberries.

Using a vegetable peeler, pare the chocolate into shavings on to a piece of parchment as the less you touch them the better.

Spoon into individual serving glasses and scatter over the chocolate shavings to serve.

Gooey chocolate brownies

These intensely chocolatey brownies are addictive. Be careful not to overcook them – they should have a nice crust while still retaining some of their gooeyness. A few minutes too long and you'll end up with a chocolate cake!

Serves 8

INGREDIENTS

150g butter

150g dark chocolate, roughly chopped (at least 70% cocoa solids)

3 large eggs

150g demerara sugar

seeds of ½ vanilla pod or ½ teaspoon vanilla extract

4 tablespoons plain flour*

pinch of sea salt

75g pecan nuts, chopped

16 tablespoons crème fraiche

*Option: 100% sieved spelt flour can be used as an alternative

METHOD

Preheat the oven to 190C/375°F and line a 20cm square baking tin with parchment paper. Melt the butter in a pan over a low heat or in the microwave. Add the chocolate and then remove it from the heat and leave the chocolate to melt.

Put the eggs, sugar and vanilla in a bowl and using a hand-held electric mixer whisk them up until light and frothy. Add the flour and gently whisk this all together along with the salt. Then pour in the chocolate and butter mixture along with the pecan nuts, gently folding it all in until well combined. Pour all of this into the prepared tin.

Pop it in the oven for 22–25 minutes until the top is crusty but there is still a nice wobble in the middle. Once baked, remove from the oven and leave to cool down a little before cutting it into eight even-sized squares, which will crack as you cut them. Serve warm with a dollop of crème fraiche, if liked.

STOCKS

Chicken stock (IBS-friendly)

A good chicken stock should have decent body, along with a mild savoury flavour that enhances, rather than competes with, the sauces, casseroles and soup bases you make with it. This version has a lovely chicken flavour with more body than water which means it should become more like a jelly when chilled. It freezes well or can be stored in the fridge for up to three days.

Makes about 1.2 litres

INGREDIENTS

1 large raw or cooked chicken carcass, skin and fat removed and bones chopped

1.7 litres of water

2 leeks, chopped (green parts only)

3 carrots, chopped

2 celery sticks, chopped

1 fresh thyme sprig

1 bay leaf

handful of fresh parsley stalks

1 teaspoon black peppercorns

METHOD

If using a raw chicken carcass, preheat the oven to 220C/450°F. Roast the carcass in a tin for about 40 minutes, until golden. This helps draw out the flavour. Drain the stock through a colander to get rid of any excess fat, then chop up the carcass.

Place the chopped-up chicken carcass in a large pan and cover with 1.8 litres of cold water. Bring to the boil, then skim off any fat and scum from the surface. Reduce the heat to a simmer and tip in all the remaining ingredients.

Simmer gently for another 1½ hours, skimming occasionally and topping up with water as necessary. If you simmer it too vigorously it will become cloudy. Taste regularly to check the flavour. When you're happy with it, remove from the heat and pass through a sieve. Leave to cool and remove any fat that settles on the top, then store in the fridge in a plastic jug tightly covered with clingfilm and use as required.

Vegetable stock (IBS-friendly)

This is a great vegetable stock as the flavour is wonderfully intense from the addition of the aromatic spices. Using it in your soups and stews will add a depth of flavour and goodness that can't be replaced with a stock cube. Like the chicken stock it freezes well or can be kept in the fridge for up to one week.

Makes 1.2 litres

INGREDIENTS

2 leeks, finely chopped (green part only)

1.7 litres water

3 carrots, diced

2 celery sticks, finely chopped

1 fennel bulb, trimmed and diced

100ml dry white wine

1 fresh thyme sprig

1 bay leaf

1 teaspoon black peppercorns

1 teaspoon coriander seeds

1 star anise

pinch of sea salt

METHOD

Place all the ingredients in a large pan and cover with 1.7 litres of cold water. Cover with a lid and bring to a simmer, then remove the lid and cook for another 45 minutes, until the vegetables are tender.

Either set aside to marinate for two days in the fridge, or if you're short of time, strain through a fine mesh sieve. Taste – if you find the flavour isn't full enough, return to the pan and reduce until you are happy with it. Leave to cool completely and transfer to a plastic/glass jug and cover tightly with clingfilm, then store in the fridge until needed. Use as required.

Suitable shop-bought stocks

There are a range of commercially ready-made stock cubes, sauces and gravies that are certified low FODMAP and are onion, leek and garlic free. These are all gut-friendly and minimize gut symptoms. The brands and availability vary from country to country.

Label-reading tip

Reading labels is important. The ingredient present in the largest quantity is first on the ingredient list. The lower the ingredient is on the list, the smaller the quantity the product contains.

The FLAT Gut Diet suitable snacks and treats

Savoury gut-friendly snacks

Nuts, seeds and crisps
One small handful of nuts or seeds – brazil, chestnuts, macadamias, peanuts, pecans, pine nuts, poppy seeds, pumpkin seeds, sesame seeds, sunflower seeds and walnuts
Ready salted crisps, salt and vinegar crisps, salted rice cakes
Kale crisps

Crackers and biscuits
Take from daily fructan allowance

Biscuits and snacks	Fructan points
Plain biscuit, 2–3 biscuits	1
Shortbread, 2–3 biscuits/fingers	1
Rice cake, 2–3	1
Cracker plain/wholemeal, 2–3 crackers	1

Toppings sweet and savoury	
Nut butters	almond, hazelnut, peanut butter
Cheeses	Brie, bocconcini, camembert, cheddar, edam, gouda, mozzarella, parmesan, Pecorino, soy, swiss, blue vein cheeses
Fish	smoked salmon, tuna, sardines, mackerel
Salads	lettuce, tomato, cucumber, peppers, olives, radish
Preserves	jam, marmalade, golden, maple syrup

Dairy foods

Kefir (150ml) or live yoghurt (100g) or Greek yoghurt (150g) or lactose-free, almond, coconut, rice yoghurts (check only contains suitable fruits). Add some chopped nuts from nut snack list.

Latte almond, rice and lactose-free milk

Cheeseboard, IBS-friendly with some grapes from fructose point allowance

Check the lactose table and take these from your daily lactose points.

Fruits	
Clementine, mandarin, kiwi	2
Strawberry, grapes – black, red, green	10 small
Blueberry, cranberries, raspberries	15
Banana, dragon fruit, orange	1
Grapefruit, pomegranate	½ average sized

Sweet treats – a little of what you fancy!

Treat yourself in small quantities. If watching your weight, limit these treats to small portions at the weekend.

Sweet treats

FLAT Gut muesli with yoghurt 100g

Gooey brownies

Oatmeal raisin cookies or oaty flapjacks

FLAT Gut breakfast bar

Dark chocolate

FLAT Gut Diet infused oils

These infused oils can be used to flavour and tenderize meat and fish prior to cooking to help keep them moist. However, remember not to overdo them or baste them during cooking, as this is what causes a flare-up, resulting in that blackened-food look that we all want to avoid.

Citrus oil

Delicious with chicken, fish or even lamb.
Makes 300ml

INGREDIENTS

1 small orange

1 lemon

1 large lime

300 ml olive oil

½ teaspoon black peppercorns

METHOD

Thinly pare the rind off the orange, lemon and lime and place in a small pan with the oil and peppercorns. Heat gently for about five minutes, then remove from the heat and pour into a heatproof glass bottle. Leave to cool completely before closing up the bottle, and use in recipes as required.

Herb oil

This oil really goes with just about anything you want to use it with.
Makes about 100ml

INGREDIENTS

100g fresh soft herbs (such as flat-leaf parsley, chives and basil)

100ml olive oil

pinch of sea salt

METHOD

Pick the leaves from the herbs and place in a mini blender, discarding the stalks. Add the olive oil and a pinch of salt and blend for five minutes, until completely smooth. Pass the herb oil through a fine mesh sieve into a jug, then transfer to a squeezy bottle and use as required. This keeps very well in the fridge for up to three weeks.

Chilli oil

I like to use this as a garnish for any dish that could use a bit of spice, but it's also fantastic brushed on seafood kebabs or prawns before cooking.

Makes about 600ml

INGREDIENTS

600ml olive oil

1 large red chilli, split in half

1 lemongrass stalk

20g fresh root ginger, sliced but not peeled

METHOD

Gently warm the oil in a heavy-based pan, but do not allow it to boil. Bring to a gentle simmer, then add the chilli, lemongrass and ginger. Continue to simmer very gently for 20–30 minutes, until the flavours are well infused. It's important not to allow it to boil at any stage. Allow to cool, and pour into a squeezy bottle, leaving the bits in as the flavours will continue to infuse. Use as required. This keeps very well in the fridge for up to three weeks in the squeezy bottle.

THE FLAT GUT DIET BONUS RECIPES

When you have got confidence in the FLAT Gut Diet and have mastered the points system, try branching out further with your meal choices, and adding more variety, with these tasty, delicious recipes.

Chicken & parmesan salad with creamy basil dressing

This is a healthy, gut-friendly alternative to your traditional Caesar salad.

Serves 2

INGREDIENTS

4 tablespoons natural yoghurt

1 teaspoon white wine vinegar

juice of ½ lemon

1 teaspoon maple syrup

¼ teaspoon prepared English mustard

small handful fresh basil leaves, shredded

1 cos lettuce or 3 little gem lettuces, separated into leaves

15g freshly grated Parmesan, extra to garnish

100g plum cherry tomatoes, halved

200g cooked roast chicken, cut into bite-sized pieces

1 tablespoon sunflower seeds

sea salt and freshly ground black pepper

METHOD

To make the dressing, place the yoghurt, vinegar, lemon juice, maple syrup, mustard and basil in a bowl and whisk until smooth. Season with a pinch of salt and plenty of pepper.

To serve, tear the lettuce leaves into bite-sized pieces and place in a large bowl. Add enough of the dressing to just coat (not drown) the leaves, then fold in the Parmesan. Toss to combine, then transfer to wide-rimmed serving bowls, scatter over the cherry tomatoes, chicken and sunflower seeds. Garnish with a little extra Parmesan.

Points per portion			
Fibre	Fructans	Fructose	Lactose
4	0	0	0.5

Speedy tomato & basil soup

This clever soup has a wonderfully intense tomato flavour and doesn't even need stock! Try to buy tins of tomatoes that are marked as Italian or they can end up being watery and disappointing in flavour.

Serves 4–6

INGREDIENTS

2 tablespoons olive oil

1 red pepper, seeded and finely chopped

1 celery stick, finely chopped

4 x 400g tins of Italian chopped tomatoes

good pinch of caster sugar

handful fresh basil leaves, plus extra sprigs to serve

sea salt and freshly ground black pepper

METHOD

Heat the oil in a large pan set over a low heat. Add the red pepper and celery and cook for about ten minutes stirring occasionally, until they are soft and lightly golden.

Stir in the tomatoes and sugar and season with a pinch of salt and plenty of pepper. Bring to a simmer and cook gently for 15–20 minutes, stirring occasionally, until the flavours are well combined. Tear in the basil and then blitz the soup with a hand-held blender.

Ladle the soup into warmed bowls and add a sprig of fresh basil to each one to serve.

Points per portion			
Fibre	Fructans	Fructose	Lactose
3	0	0	0

Roast leg of lamb with potato gratin & wilted spinach

This dish is hassle-free so it's perfect for a dinner party or Sunday lunch. The leg of lamb is roasted on a rack sitting directly over the layered-up potatoes so that they collect and absorb all the flavours and juices released from the meat as it cooks.

Serves 6

INGREDIENTS

1.5kg potatoes, peeled and thinly sliced

4 fresh thyme sprigs, leaves stripped

400 ml chicken stock (IBS-friendly, see 'Stocks', above)

1.75kg leg of lamb

2 fresh rosemary sprigs

30g butter

500g fresh spinach leaves, tough stalks removed

pinch of ground nutmeg (optional)

sea salt and freshly ground black pepper

METHOD

Preheat the oven to 200C/400°F. To make the potato gratin, layer the potatoes, and thyme leaves in a roasting tin large enough to sit neatly underneath the leg of lamb. Season each layer as you go with a pinch of salt and plenty of pepper and finish with an attractive overlapping layer of the potatoes. Pour over the stock and set to one side.

Using a sharp knife, make small incisions all over the lamb and press a tiny sprig of the rosemary deep into each one. Weigh the joint and allow 20 minutes per 450g plus 20 minutes (add a further 20 minutes for well done) then place the lamb carefully on the rack on top of the potatoes in the roasting tin. Roast for 20 minutes and then reduce the oven temperature to 180C/350°F and roast for one hour and 40 minutes for a leg of lamb this size.

Transfer the lamb to a carving platter and cover loosely with foil, then leave to rest for 15 minutes, keeping the potato gratin warm.

To prepare the spinach, heat a pan over a medium heat and add the butter. Once it has stopped foaming, quickly sauté the spinach with the nutmeg, if using, until it is soft and wilted. Season with a pinch of salt and plenty of pepper and then drain briefly in a sieve or on kitchen paper to remove the excess moisture. Return to the pan and keep warm.

To serve, carve the lamb into slices and arrange on warmed serving plates with the potato gratin and wilted spinach.

Points per portion			
Fibre	Fructans	Fructose	Lactose
6	0	0	0

Parma wrapped monkfish with crispy new potatoes & kale

This dish can all be prepared in advance and the monkfish cooked just before you want to serve it. Make sure your fishmonger gives you the monkfish well-trimmed as the membrane can be quite tricky to remove.

Serves 4

INGREDIENTS

4 x 175g monkfish tail fillets

50g roasted red peppers, thinly sliced (from a jar is fine)

2 tablespoons shredded fresh basil

8 thin slices Parma ham, about 175g in total

2 tablespoon extra-virgin olive oil

knob of butter

500g baby new potatoes, halved or quartered into bite-sized pieces

350g curly kale, stems removed and leaves roughly chopped

salt and freshly ground black pepper

METHOD

Preheat the oven to 190C/375°F. Cut a slit in the side of each piece of monkfish to create a pocket. Open each pocket out a little and arrange the roasted red peppers and basil inside, then season with a pinch of salt and plenty of pepper. Close over, sandwiching in the pepper mixture. Lay two Parma ham slices out on the work surface, slightly overlapping, then place one monkfish fillet on top and wrap in the Parma ham to enclose the monkfish completely. Repeat with the remaining fillets.

Heat a tablespoon of the olive oil in a large ovenproof frying pan and sear the monkfish pieces all over for three to four minutes until the Parma ham is crisp and golden. Transfer to the oven and roast for another ten to 12 minutes until firm to the touch and just cooked through.

Meanwhile, place the potatoes in a pan of cold water over a high heat and add a pinch of salt. Bring to the boil and cook for five minutes until partially cooked. Drain and set aside.

Heat the remaining oil with the butter in a large non-stick frying pan over a medium heat. Add the potatoes, then season with a pinch of salt and plenty of pepper, and sauté for eight to ten minutes, stirring occasionally, until lightly golden and tender.

Increase the heat to high, add the kale and sauté for another two to three minutes, until the kale is tender and just crisp around the edges. Arrange the Parma-wrapped monkfish on warmed plates with the crispy new potatoes and kale alongside to serve.

Points per portion			
Fibre	Fructans	Fructose	Lactose
6	0	0	0

Luxury fish pie with cheesy potato rosti

This is the ultimate fish pie – a smoky, creamy base of the pie topped with a crispy cheesy potato topping. For best results, choose fish fillets from the centre cut, as they will cook more evenly. Don't be afraid to experiment with the combination of fish, but never use more than half the quantity of smoked fish or its flavour will be overpowering.

Serves 4–6

INGREDIENTS

450g floury potatoes, such as Maris Piper

50g mature Cheddar cheese

250g skinless and boneless smoked haddock or cod fillet (undyed if possible)

250g skinless and boneless salmon fillet

100g raw peeled prawns

400ml full-fat milk or lactose-free milk

1 celery stick, finely chopped

1 bay leaf

1 teaspoon black peppercorns

75g butter, softened

50g 100% spelt plain flour

2 teaspoons Dijon mustard

2 tablespoons chopped fresh dill

sea salt and freshly ground black pepper

tossed crisp green salad, to serve (optional)

METHOD

Preheat the oven to 200C/400°F. Peel the potatoes, then grate using the coarse side of a box grater. Place the grated potatoes in a bowl of cold water to stop them from turning brown. Grate the cheese and set aside separately.

Place the smoked haddock or cod, salmon and prawns in a pan, pour over the milk and drop in the celery, bay leaf and peppercorns. Cook over a low heat for two to three minutes until just tender. Using a fish slice, transfer the fish and prawns to a pie dish and then pass the poaching milk through a sieve, discarding the flavourings.

Melt 50g of the butter in a pan over a medium heat. Add the flour and cook for one minute, stirring constantly. Gradually whisk in the poaching milk until you have a thick smooth sauce. Season with a pinch of salt and plenty of pepper and stir in the mustard and half the dill. Use to cover the fish and prawns in an even layer.

Place the grated potatoes in a clean tea towel and twist out all the excess liquid until they are completely dry. Put in a bowl and mix with the rest of the dill and the cheese. Season with a pinch of salt, then scatter over the sauce-covered fish. Melt the remaining butter in a pan and brush over the top of the pie. Bake for 25–30 minutes or until crisp and golden brown. Serve straight to the table with the tossed salad, if liked.

Points per portion			
Fibre	Fructans	Fructose	Lactose
4	0	0	1

Oatmeal raisin cookies

Everyone needs a good chewy cookie recipe, and this one is great as it uses plenty of oats which gives them a lovely texture and flavour. They will keep up to three days if stored in an airtight container, although it can be hard to resist them that long!

Makes 10

INGREDIENTS

100g butter (at room temperature)

100g demerara sugar

1 medium egg

seeds of 1 vanilla pod or 2 teaspoons vanilla extract

70g raisins

50g 100% spelt self-raising flour

130g porridge oats

pinch of sea salt

METHOD

Preheat the oven to 180C/350°F and line two baking sheets with parchment paper. Put the butter and the sugar into a bowl and cream it together. You can do this with a wooden spoon, but if you have a hand-held electric whisk then that will make things much easier and quicker. Add the egg and vanilla and beat well, and then stir in the raisins, flour and oats along with the salt.

When you have added the flour bring the mixture together using as few stirs as possible so that the dough does not get too tough. Then divide the mixture into 10 equal-sized blobs, (each about 50g) placing them on the prepared baking sheets as you go and leaving plenty of space for them to spread out. Put in the oven and cook for eight to ten minutes. The trick is to pull them out of the oven before they are super firm – you want them to come out just slightly under-baked so that when they cool, they are still chewy. Once they are ready, remove them from the oven and leave to cool a little. These are also delicious, almost irresistible, when warm.

Points per portion			
Fibre	Fructans	Fructose	Lactose
2	0	0	0

The FLAT Gut Diet soothing tea

If you are feeling a little bloated this is our favourite soothing tea remedy. We love to use this at the weekends and days off; it's great for hydration and digestive comfort.

Makes 1 teapot

INGREDIENTS

boiled water

handful cup of fresh mint leaves

small handful of fresh mint leaves in a teapot

add 2 peppermint tea bags

juice ½ lemon optional

METHOD

Boil a kettle of water with enough to fill your teapot. In the meantime, add the peppermint tea bags and a small handful of fresh leaves to the teapot. Once boiled, add the water to the pot and allow it to steep for five to ten minutes.

Add a few mint leaves to your cup (optional) and pour in the tea once infused. You can add a little freshly squeezed lemon if you wish.

22

The FLAT Gut Diet Steps 2 & 3 – Tolerance and Sustain

Now that you have completed the Step 1 – Simplify 4-week plan, you're ready to challenge your gut tolerance to the FLAT Gut food groups.

Step 2 – How to start the FLAT Gut Diet tolerance testing

- **8–10 weeks duration:** You'll gradually increase your daily intake of the trigger foods with the aim of identifying the food groups to which you are most sensitive.
- **One at a time:** You choose one FLAT Gut food and increase the points you're eating, according to the plan, while keeping the rest of your daily points the same.
- **You choose the order:** In the Tolerance Testing, you may choose the food group to test in any order you prefer. The order that we suggest is as follows: Fibre, Lactose, Fructans, Alliums, Fructose and GOS.
- **You choose the time:** Choose a time that's best for you to carry out the tolerance test. Most people find weekends are the best.

Finding your food triggers and testing your tolerance

TIPS FOR STARTING YOUR FOOD TOLERANCE TESTING

- We suggest carrying out each test over a weekend when you have no plans made, so that you are in the comfort of your own home. This will reduce any anxiety regarding the onset of symptoms when you are busy or away from home.

- If you work weekends, choose a time when you are at home for a period of approximately 48 hours.
- If you have any upcoming family events (weddings, communions, confirmations, holidays) or sporting events (hike, marathon, cycle), you may benefit from returning to Step 1 for one week prior to the event.
- If you tolerate a particular food on the tolerance test, then you can continue to incorporate it into your daily diet.
- If you experience symptoms at any point during these food tests, stop the test, as you have passed your tolerance threshold for that particular food. You could try again the next day with an even smaller amount if you wish. Wait until the following week to start the next tolerance test.
- Often it will take several days of eating a particular food to reach your level of tolerance threshold. At that point you will start to experience symptoms.

The FLAT Gut Diet Step 2 – Tolerance testing		
Fibre	20→25	Points per day
Lactose	2/3→3/4	Points per day
Fructans	2/3→3/4	Points per day
Alliums	0→1/2	Points per day
Fructose	2/3→3/4	Points per day
Polyols	0→1	Points per day
GOS	0→1	Points per day

Food tolerance testing: food and symptom diary

This is a food and symptom diary that we advise you to use when testing your tolerance for the different food groups. Fill in the food group you are testing, e.g. lactose, and the food chosen, e.g. yoghurt. Then document your gut symptoms, if any, associated with eating this food.

Food tolerance testing – food and symptom diary

Symptom record	
Day 1	
Food group testing tolerance	*e.g. lactose*
Food being tested, quantity and when eaten	*e.g. yoghurt 100g, 3pm*
Details of any symptoms experienced	*e.g. mild bloating and increased wind after lunch*
Day 2	
Food being tested, quantity and when eaten	*e.g. yoghurt 100g, 7pm*
Details of any symptoms experienced	*e.g. mild bloating and increased wind after evening meal, loose bowel movement 10pm*
Day 3 – no test food to be eaten	
Details of any symptoms experienced	*e.g. none/abdominal bloating mid-morning*
Subsequent days	
Details of any symptoms experienced	*e.g. none*

Week 1: fibre tolerance

- As a first step, we would like you to increase your intake of fibre by 5 points to 25 points per day. We advise you to keep your fibre intake at this level while you undertake the other tolerance tests.
- Vegetarians and vegans may already be tolerating in excess of 25 points of fibre daily, as their diets are normally very high in fibre. However, in these cases we recommend a maximum fibre intake of 25–30g fibre daily during their tolerance testing.
- When all tests are completed, you can test your fibre tolerance further by increasing your intake to 30–35 grams per day. However, those with IBS-D may not be able to tolerate this amount of fibre, while those with IBS-C may tolerate a greater amount of daily dietary fibre. The aim is to find the best amount of daily fibre for you.

Fibre tolerance
Introduce 5 extra fibre points increasing from 20→25 points per day
Do this for 2 consecutive days
Choose foods from the fibre points table in Chapter 20 Fibre 5-point examples 5 points = 1 tablespoon (15g) chia or linseed/flaxseed or 1 extra portion of both salad and vegetables daily
We recommend leaving extra fibre from fruit until the fructose tolerance test.

Week 2: lactose tolerance

Dairy is an important part of most people's diets, providing essential calcium, vitamins and proteins for bone health and overall health. Unless you are lactose intolerant, it is all about the quantity of lactose that you can tolerate. Remember, there are plenty of naturally low lactose foods such as cheese, butter and cream that you can choose.

Lactose tolerance
Introduce 1 extra lactose point increasing from 2→3 points per day
Do this for two consecutive days
Choose from the lactose points table in Chapter 20 Lactose 1-point examples: 1 point = extra 100ml cow's milk or 100g extra yoghurt per day

Week 3: fructan tolerance

It is all about quantity! As wheat-containing foods are the most common source of fructans, we'd now like you to test your tolerance to different wheat-containing foods.

Fructan tolerance
Introduce 1 extra fructan point increasing from 2/3→3/4 points per day
Do this for two consecutive days
Choose from the fructan table in Chapter 20 Fructan 1-point examples: 1 point = 1 extra slice of bread or 20g pasta dry weight (if eating a pasta dish, for this test you eat 80g – 4 points – rather than your previous 60g – 3 points in Step 1)
Note: fresh egg pasta is better tolerated initially than dried pasta

Week 4: allium tolerance testing – onion

Onions contain potent fructans, which is why we consider them separately from the other fructan-containing foods, such as wheat. Test your tolerance by introducing onion, and gradually increasing the amount over two days.

Alliums I: onion tolerance
Introduce onion starting at 1 point per day
If tolerated, increase from 1→2 points on day 2
Do this over two consecutive days
Example of onion 'allium' points: 1 point = ½ medium or ¼ large onion = 40g 2 points = 1 medium or ½ large onion = 80g
How to do this in practice – add the onion to any meal of your choice; for example a salad, or omelette or other cooked dish.

Week 5: allium tolerance testing – garlic

Just like onions, garlic also contains potent fructans. We're using more of this in our cooking than in the past, which is why nowadays, garlic is a very significant food trigger of both IBS and FD. Test your tolerance by introducing garlic and gradually increasing the amount over two days.

Alliums II: garlic tolerance
Introduce garlic starting at 1 point per day
If tolerated, increase from 1→2 points on day 2
Do this over 2 consecutive days
Example of garlic 'allium' points: 1 point = ½ garlic clove 2 points = 1 garlic clove
How to do this in practice – add the garlic to any meal of your choice where you can measure the portion of garlic that you will eat; for example an omelette, making some garlic butter garnish, or garlic bread (the bread must be within your daily fructan allowance).

Week 6: fructose tolerance

We all love fruit, so it's good to try to increase your intake at this point. Remember to spread your intake over the day, as a high fructose load consumed in one sitting can cause troublesome gut symptoms. You may be surprised at how you tolerate fruit better when you eat it in smaller quantities and choose fruits that are low in fructose.

Fructose tolerance
Introduce 1 extra fructose point increasing from 3→4 points per day
Do this over two consecutive days
Choose from the fructose points table in Chapter 20
Fructose 1-point examples 1 point = 10 small grapes or 10 small strawberries

Weeks 7–10: à la carte tolerance testing: you choose

You've just successfully navigated through four weeks of Simplify and another seven weeks of various tolerance tests, and by now there may be some foods in the 'avoid' sections that you are missing or even craving. This is your opportunity to decide what foods you would like to reintroduce to try to find your tolerance level. If it's something you love, we're crossing our fingers that it goes well.

In Chapter 20 we showed you a lot of tables (yes, a *lot*) to guide your food choices, and we also recommended that you avoid certain foods altogether; these included:

- legumes (high in GOS).
- certain fruits and vegetables that are high in polyols – for example cauliflower, mushrooms and stone fruits.
- certain fruits/foods that are high in fructose – such as honey and mango.
- alcohol – you may wish to enjoy an alcoholic drink. See our Alcohol Chart in Chapter 20 for further details.
- you might also like to try onions and garlic together, to see if you can combine these, even in small amounts, in your cooking.

Now is your opportunity to test your tolerance level to these foods and we're including a couple of additional charts to give you some guidance on how to do this.

THE LEGUME POINTS CHART

The FLAT Gut Diet STEP 2 – Tolerance	Legume points
Small amounts of legumes can be trialled	
Portion: 2 tablespoons	1
Chickpeas (canned), lentils (canned), lima beans (boiled), mixed beans (canned), mung beans, adzuki beans (boiled/canned), butter beans (canned)	
Portion: 1 tablespoon	1
Borlotti beans (canned), fava beans, green/red lentils (dried & boiled), hummus, red kidney beans (boiled/canned), soya beans (boiled), split peas (boiled)	

THE POLYOL POINTS CHART

Polyol foods Portion control required	Portion size	Polyol points The FLAT Gut Diet Step 2 – Tolerance
Fresh fruit – apple, apricot, blackberries, cherry, coconut**, nectarine, peach, pear, plum	40g portion	1
Dried fruit – apricot and prune	20g portion	1
Vegetables* – avocado*, cauliflower, celery*, corn on the cob, fennel bulb, mangetout*, mushroom (button, enoki, portobello, shiitake), sweet potato	40g portion	1

*some are allowed in small quantities in the fibre vegetable section
**coconut dessicated tolerated in small amounts and used in some of FLAT Gut recipes

À la carte tolerance testing: you choose
Start with 1 point of the food in question
Do this for two consecutive days
Choose from the polyol or legume table

Examples of suitable quantities:
Legumes: 1 point = 2 tablespoons of canned chickpeas
             = 1 tablespoon of red kidney beans (canned/boiled)
Polyols: 1 point = 40g avocado (approx. ¼ large avocado)
             = 40g of mushrooms

If 1 point is tolerated, then increase to 2 points on the following day

> Alliums combo:
> If combining onions and garlic, start at your tolerance level for each and combine: if you tolerated 1 point garlic and 1 point onion – add the two together.

We recommend that you continue to test your tolerance of foods and especially of any new foods that you want to try. Now that you know how to do it, keep experimenting and testing your tolerance to other high fructose fruits like honey and apples, or to polyols like cauliflower and so forth. You are now an expert on your own gut health and on how to eat with confidence in all situations.

Step 1 – Simplify and Step 2 – Tolerance: completed

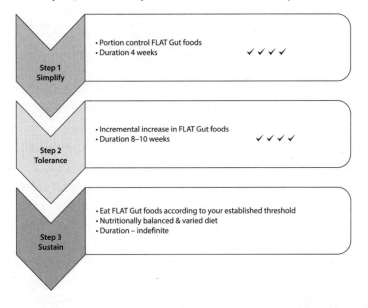

Once you have completed your 12–14 week journey on the FLAT Gut Diet, you should have noticed a marked reduction in your digestive symptoms. You will also have a greater knowledge of your trigger foods and have a real gut instinct for the quantity of these foods that *you* can tolerate without developing symptoms.

Don't forget that all aspects of everyday life including travel, work, sleep, exercise and stressors of all kinds, also play their part in the orchestra of the gut–brain axis and in your gut symptoms. This is something you'll need to keep working on long term so as to optimize your gut health and to enhance your overall wellbeing.

Step 3 – Sustain

This final step is the beginning of your long-term journey to sustain and maintain a nutritionally balanced, diversified diet. You know how to limit your trigger foods so that you have control of your life and your symptoms, rather than your symptoms controlling your life and you.

The FLAT Gut Diet Step 3 – Sustain

General guidelines for ranges tolerated	
Fibre	25–35 points per day
Fructans	3–4 points per day
Fructose	3–4 points per day
Lactose	3–4 points per day
Alliums	1–2 points per day Individual Tolerance Levels
Total	Wellbeing – mind and body

Personalized nutrition-based guidelines are the ultimate goal, with you in control of what you eat. Your gut, its unique GM and your gut–brain axis all play a role in determining your personalized food tolerances.

You can now begin to include foods in the 'Avoid' sections in very small quantities at this stage. However, as before, you will need to assess your symptoms to see if you can tolerate them. Some individuals can eat 40g of fibre per day with no side effects while others can barely tolerate 20g of fibre per day. Remember that one fibre point equals one gram of fibre.

On this journey we have asked you to test and track your individual responses to the FLAT Gut food factors and by now we hope that you are your own digestive expert.

We want you to be able to enjoy your food, to enjoy variety in your diet – variety in type, colour and quantities. It is our mission to help you succeed in managing your IBS symptoms with this knowledge and guidance.

More than just a diet

Having treated many thousands of patients, we have both seen the great improvements that can be achieved with the right dietary approach. But it's important that you don't forget the 'T'.

23

'T' for TEAMS and Total mind and body health

Once you have followed the FLAT Gut dietary programme, you will have become an expert at tracking your body's response to the foods you eat, and should have a good understanding of your tolerance level for the main trigger foods of your unpleasant gut symptoms.

We've mentioned the TEAMS factors, that are so important as part of your overall management of both IBS and FD (Chapters 14 & 16). In this chapter, we want to give you some specific targets and advice to help you actively engage with the 'T', hand in hand with your dietary solutions.

Recap: what is 'Total mind and body health'?

T   Total mind and body health
E   Exercise
A   Alcohol
M   Mental Health
S   Sleep

Exercise

We would like you to incorporate regular exercise into your weekly regime. This need not be extremely strenuous, high-octane exercise; gentler exercise is also beneficial.

> #### The Gut Experts' exercise tips
>
> - Walk or take some other form of exercise for 30 minutes at least five times per week.
> - Make exercise fun; do something you enjoy.
> - Walk instead of driving or taking public transport where possible.

(Continued)

- Use stairs rather than lifts where possible.
- Find a friend to exercise with or join a group or class.
- Yoga, Qi-gong and Tai chi are very effective at connecting and relaxing the gut–brain axis.
- Set weekly exercise goals.

Alcohol

We spoke about the many effects that alcohol can have on our digestive system and gut symptoms. In general, the less you drink, the happier your gut will be. If you do enjoy an alcoholic drink, have a look at our recommendations below, as adhering to these will help minimize your gut symptoms.

The Gut Experts' advice on alcohol consumption

- Stay within recommended low-risk guidelines for alcohol consumption; the guidelines vary a little from country to country in terms of size of a unit or standard drink. But all the guidelines work out similarly at around 110g alcohol per week for women, spread out over the week.
- Have at least two to three alcohol-free days each week.
- Drink with or after food, not on an empty tummy.
- Try to avoid foods that trigger IBS if drinking alcohol.
- If you have FD, keep alcohol to a minimum.
- Alternate alcoholic drinks with water as alcohol dehydrates the body.
- Wine (red, rosé and white), vodka, gin and whiskey are best tolerated – but obviously moderation is key.
- Cocktails containing fruit juices high in fructose are known to trigger IBS symptoms.
- Beer tends to cause wind/gas and bloating.
- Carbonated mixers in large quantities can cause gas and bloating.
- Avoid rum, dessert wine, cider, port and sherry.

Mental health

Your mental health must be actively nurtured. More importantly, if you have symptoms of anxiety or depression, it is extremely important that you speak to your doctor or an appropriate healthcare professional. Significant ongoing mental health issues can worsen your IBS or FD symptoms, and can prevent you from responding to dietary measures. Both medications, and non-pharmacological treatments such as cognitive behavioural therapy (CBT) and gut-directed hypnotherapy, have been shown to be effective treatments for IBS.[86–88] Yoga has also proved an effective treatment for some people and is also something that you could explore.

The Gut Experts' advice for mental health

- Speak to a healthcare professional if you are experiencing significant mental health problems. She/he will help you to decide on the best course of action in terms of medication or other therapy.
- Consider CBT if stress/anxiety is having an impact on your digestive symptoms.
- Yoga, mindfulness or hypnotherapy can all be helpful. See if a class or group is available near you and sign up.
- There are online resources and apps available.
- Daily practice of ten minutes every day for some meditation/mindful exercises. Have your daily practice in a place that is quiet, warm and peaceful.
- Keep a gratitude list – write down three things you are grateful for each day. It's a simple thing but is very effective in focusing the mind on the positive.

Sleep

Shakespeare could have taught us all a thing or two about sleep; 'O sleep, O gentle sleep, nature's soft nurse', or 'sleep that puts each day to rest', suggests that they knew exactly how important sleep was back then. Sleep became hugely undervalued in Western society in recent years, with many people believing or being led to believe that time spent sleeping was time wasted. It has seemed as if getting only five or six hours or even less of sleep is something to be proud of – it isn't. People with IBS also more commonly report poor sleep as being a problem. This can lead to increased anxiety, fatigue and even lack of clear thought or poor cognitive function, particularly if occurring over successive days. This can all lead to a worsening of gut symptoms as the gut–brain vicious cycle kicks in. Therefore, it's important that we try to optimize our sleep and our so-called 'sleep hygiene'. Lack of sleep can also affect us on the weighing scales: a 2016 study found that 'sleep-deprived people consumed an average of 385 extra calories per day' – that's the equivalent of about five slices of bread.

The Gut Experts' sleep advice

- Most adults need seven to eight hours a night. Try to increase your 'sleep opportunity' by allowing yourself enough time in bed to achieve this sleep target, i.e. if you are aiming for seven hours sleep, you may actually need to allow eight hours in bed to give yourself enough time to fall asleep etc.
- Try to set a regular bedtime and wake-up time as this reinforces your circadian rhythm; essentially your internal clock that controls the desire for sleep and wakefulness. Experts suggest that we should set a 'to-bed' alarm as well as a wake-up alarm.
- Avoid using blue light devices e.g. smartphones or tablets an hour before bedtime – studies have shown that blue light suppresses the production of the hormone melatonin which aids sleep. If you must use these devices, consider blue light-blocking glasses.
- Allow yourself 60–90 minutes to wind down before bedtime; take a warm bath, read a book, journal, do some gentle yoga or stretching. Vigorous exercise close to bedtime can disrupt sleep.
- Avoid caffeine late in the day (coffee, tea or dark chocolate). If you're having trouble sleeping, it may be helpful to avoid caffeine after 4pm. Enjoy herbal teas instead.
- Avoid eating two hours before bedtime.
- Alcohol reduces sleep quality, and should be kept to a minimum if you are having trouble sleeping.
- Don't work from your bed; your bedroom should be a space for rest, some light reading or sex.
- Use an eye mask and ear plugs if you are having trouble sleeping. Try guided sleep meditations – there are lots of good apps available now.
- If you find that these measures do not help your sleep, speak to your doctor.

The Gut Experts' caffeine tips

- Caffeine has a 'half-life' of five hours in our body. This means that when we eat or drink something containing caffeine, half of it remains in our bodies five hours later and will gradually be cleared from our system.
- The stimulatory effect of caffeine occurs with 15–30 minutes after consumption.
- The average caffeine content of a cup of brewed coffee is 111mg, instant coffee is 78mg, tea is 44mg and caffeine-containing soft drinks have approximately 30mg per 500ml.

Diet and TEAMS together

By combining the FLAT Gut Diet and a TEAMS approach, you have all the tools you need to minimize your problematic gut symptoms. By reading this book, you also have a broad knowledge and understanding of how the female digestive system functions, no matter what stage of life you are at. We are not saying that it will be easy, but if you have IBS or FD, achieving better control of your symptoms is worth it.

And to conclude....

Mission complete?

In our introduction, we stated that it was our mission to 'Improve quality of life through better gut health.' We hope that over the course of this book we have helped you to achieve this goal. Remember the Three 'E's:

Educate: You should have a more complete understanding of how your gut works, how symptoms are investigated, how food affects you and the importance of a balanced diet and lifestyle.

Empower: You now have the knowledge and skills to control your symptoms, rather than to have them controlling you. By following the advice in this book, you will have learned how to challenge yourself mentally and physically. It is no mean feat reading and digesting all this information and then adhering to a 12-week programme of any kind. That takes real tenacity. We also hope that you now know how to challenge the false and misleading information out there, so that you can make informed and wise decisions, based on sound scientific evidence.

Eating for health & wellbeing: The FLAT Gut Diet provides you with a roadmap to finding the balance of healthy foods that suit your individual digestive system. This dietary approach is simple to follow, does not require the elimination of many foods from your diet, and gives you control over your symptoms, while still ensuring that your diet is nutritionally complete. This diet will sustain you and help to control your symptoms over the long term. Remember that this is a marathon and not a sprint.

Almost there...

As we said at the start of this book, there is no 'cure' for IBS or functional dyspepsia, but there are many things that can be done to help manage the symptoms. Managing these conditions is a way of life. Our FLAT Gut Diet is simple (once you get the hang of it). Once you make some basic changes, it allows for plenty of diversity and variety in the long term. You become an expert really quickly. This book is just the beginning of your journey to gut health.

Still hungry for information?

We thought that it would be helpful to touch on some additional topics, which we find are of particular interest to our patients. These include brain fog, leaky gut, SIBO, probiotics, chronic fatigue, to name but a few. If you're interested, you can dip into these chapters at any time – we think that we've got some helpful information in there, to allow you to make more informed choices about your own gut health.

Section 6

FAQS – THINGS WE'RE OFTEN ASKED ABOUT

24

Do you have small intestinal bacterial overgrowth (SIBO)?

Some background facts about SIBO

- It is not normal to have large numbers of bacteria growing in the small intestine, and when this happens, it is called small intestinal bacterial overgrowth, or SIBO.
- SIBO is defined as the presence of excessive numbers of bacteria in the small intestine causing gastrointestinal symptoms.[89]
- Normally the numbers of bacteria in the small bowel are in the order of billions, but if these proliferate, they can interfere with digestive processes and cause unpleasant symptoms.
- The bacteria can proliferate anywhere along the small bowel, but tend to be more common towards the lower end.
- The bacteria and other organisms in the small intestine are hard to study (because this part of the bowel is so long and difficult to access) and as a result we have identified only a fraction of the microorganisms that normally live there, resulting in these being called the 'mysteriome'.[2]
- SIBO is very different from the 'dysbiosis' that we spoke about previously; dysbiosis refers to the wrong mix or reduced diversity of the bacteria in the colon.
- An excess of methane-producing microorganisms (strictly speaking these are like bacteria but are actually called archaea) seems to be more common in patients who are prone to constipation.

Why do bacteria in the small bowel cause problems?

- The function of the small bowel, as we explained earlier, is to digest and absorb nutrients.
- The food is broken down into small particles like sugars, simple fat molecules and the building blocks of proteins called amino acids. When there are bacteria present they interfere with the absorption of these chemicals and digest (ferment) the sugars and amino acids themselves.
- A by-product of this bacterial fermentation is gas production, which can lead to bloating and flatulence.
- Other by-products of bacterial fermentation can interfere with salt and water absorption from the bowel, leading to diarrhoea and bloating.
- Bacteria in the small bowel can also impair absorption of important micronutrients and vitamins such as B12, vitamin D and iron.

What causes SIBO?

As with many things in the human body, there are a number of reasons why someone might get SIBO.[89,90]

Causes of SIBO

No obvious cause: Sometimes there is no obvious cause at all.

Gut motility problems: Normal small bowel contractions move contents of the small bowel along like a small mountain stream, containing clear, fresh water. If flow slows down, the water becomes a bit murky and overgrown with weeds. The same thing happens in the small bowel; if the contractions weaken and the flow of contents slows down, bacteria can multiply, giving rise to SIBO. Causes of motility problems include:

- IBS
- diabetes
- neurological conditions
- medications like opioids

Reduced gastric acid: Most bacteria we ingest with food cannot survive the acid conditions of the stomach and reduced gastric acid increases the risk of SIBO. Causes include:
- proton pump medication (PPI); these are commonly prescribed for many reasons and increase the risk of SIBO twofold.
- previous stomach surgery.

Previous surgery – a number of surgeries increase the risk:
- stomach surgery
- previous bariatric surgery for weight loss
- surgery on the large bowel

What are the symptoms of SIBO?

The symptoms of SIBO can be very nonspecific but include the following:

Symptoms of SIBO

- Abdominal pain and cramps
- Bloating
- Excessive wind and flatulence – this can have a very strong odour
- Indigestion
- Diarrhoea (less commonly there can be constipation)
- Alternating bowel pattern (constipation and diarrhoea)
- Nausea
- Borborygmi (loud gurgling bowel sounds)
- Brain fog

How is SIBO diagnosed?

The hydrogen and methane breath test is the most common test used in every day clinical practice to detect SIBO. This is a simple and non-invasive investigation. A baseline breath sample is measured by blowing directly into the analyser and after this a drink is given made from either glucose or lactulose. Breath samples are then taken at regular intervals for several hours. The test is not perfect – overall accuracy is 60–70 per cent; some cases can be missed with this test (false negative test); the test can also give a false positive result, particularly if someone has a very active bowel (fast bowel transit).

IBS *or* SIBO or IBS *and* SIBO?

Looking at the list of potential symptoms caused by SIBO, the first thing you'll notice is that it looks very like the list of symptoms caused by IBS! Are some people misdiagnosed with IBS when in fact they have SIBO? Or do some people have both IBS and SIBO? So, what is the evidence for SIBO and IBS?

- Studies have shown that around 20–50 per cent of people with IBS have SIBO.[91-93] Interestingly 5–20 per cent of healthy controls (people with no digestive complaints) were also found to have a positive breath test.
- IBS patients are around four times more likely than healthy controls to have SIBO. If you have IBS, and your symptoms are not settling with first-line treatments, *you should be tested for SIBO.*
- Many different antibiotics have been tried as a possible treatment for IBS over the years and most studies show a benefit with antibiotic treatment for a significant number of patients with IBS. Not all of those studies tested for SIBO, and it is not certain whether the benefit seen was due to treatment of co-existing SIBO, or an alteration in colonic bacteria, which is also known to be altered in IBS. On average, one in nine people with IBS treated with an antibiotic will see a significant improvement in their symptoms.[94]
- Most clinicians prefer to test for SIBO and treat with antibiotics if this is positive. However, given that the breath test is not perfect, if someone has symptoms strongly suggestive of SIBO, some doctors will still consider it worthwhile to give a trial of antibiotics even if the test is negative.

Which IBS patients are more likely to have SIBO?

Factors associated with a higher risk of SIBO in IBS patients

- Being female
- Those with diarrhoea-predominant IBS (IBS-D)
- Symptoms of bloating and flatulence
- Taking a PPI medication
- Over 55 years of age
- Using opioid medications

Other conditions associated with SIBO

- Autoimmune diseases
- Fibromyalgia
- Acne rosacea
- Diabetes
- Polycystic ovary disease (PCOS)

Treatment of SIBO

The aim of all treatments for SIBO is to eradicate the bacterial overgrowth and relieve the symptoms. A number of different strategies have been tried to achieve this.[89]

1 Antibiotics

Many different antibiotics have been tried to treat SIBO. Rifaxamin is one of the most studied and has the advantage of not being absorbed from the bowel into the bloodstream, so acts locally within the gut. Rifaxamin is about 70 per cent effective, especially if hydrogen-producing bacteria are present. If there is methane production, a second antibiotic is required in combination and either Metronidazole or Neomycin are commonly used. Other antibiotics, either alone or in combination, have shown eradication success rates of 50–80 per cent.

The optimum duration of treatment required to achieve eradication is not known, but treatment should be for at least two weeks, and possibly up to four weeks.

2 Probiotics

The role of probiotics in treating SIBO is not clear at this point. One recent study, which pooled together all the data from existing research, suggested that probiotics may be effective in treating SIBO in some people, but long-term use did not appear to be effective in preventing its recurrence. This area needs further study and clearly would be of huge interest to those affected by SIBO.

3 Diet

Gut bacteria digest and ferment undigested sugars and carbohydrates and diets low in these carbohydrates, particularly highly fermentable ones, have been suggested as a potential treatment for SIBO. There is some evidence that low fermentable diets (such as a low FODMAP diet) reduce hydrogen production by gut bacteria,[89] and could be a useful addition to other forms of treatment for SIBO. As we have mentioned before, these diets are restrictive and can be low in fibre, which is not ideal in the long term.

4 Herbal remedies

Some natural substances such as thyme, sage, oregano, wormwood, berberine, grapefruit seed, garlic, ginger, rhubarb and licorice, to name but a few, are commercially available in combination as capsules. Many of these have been shown to have antibacterial properties and are commonly used in Chinese medical practice. We know that over half of people with IBS use complementary and alternative medicines, including herbal products.

An interesting study published some years ago found that a month-long course of herbal remedy was as effective as Rifaxamin antibiotic in eradicating SIBO.[95] This area would benefit from further study: are some herbal therapies more effective than others? Are some better tolerated? We know that herbal remedies can have side effects, just as prescribed medications can, so more study and regulation in this area would certainly be helpful.

Recurrent SIBO

As we said earlier, SIBO often develops in people who have an underlying condition. It's not all that surprising then, that SIBO can recur even after successful eradication. Almost 50 per cent of people diagnosed with SIBO will develop a recurrence of their symptoms within a year of treatment.[90] Ideally, any underlying cause should be treated, but this often isn't possible, and some patients will require repeated courses of antibiotics at various intervals. Most clinicians will aim to give patients as little antibiotic as possible, and to stretch out the interval between courses as much as possible. If you have been diagnosed with SIBO, you may be able to recognize when symptoms are beginning to recur and will know when to re-start an antibiotic, based on your symptoms.

Many people with recurrent SIBO will also try dietary measures such as a low carbohydrate/high protein approach, but this can be difficult in the long term. The potential role of herbal remedies in this context is intriguing.

Take-home messages

SIBO:

- Is more common than previously thought;
- Should be considered if IBS symptoms are not responding to other treatments;
- Is more common in women with IBS;
- Is more common in older women (>55 years);
- Can be a recurrent problem.

25

Why do you have brain fog, aches and pains and chronic fatigue? Do you have a leaky gut?

IBS is associated with a range of conditions, some well known, others less so. If you have IBS, you may be aware of these, you may have experienced their symptoms and you may indeed find some of the strategies and insights in this chapter very useful for understanding and managing your condition.

Leaky gut – is this real?

'Leaky gut' syndrome, which has attracted much attention in recent years, has been widely discussed in complementary/alternative medicine (CAM) circles. An internet search of the term 'leaky gut' will generate thousands of hits, implicating a leaky gut as the cause of many diverse medical conditions ranging from autism to obesity to IBS. So, what is the current state of knowledge regarding a leaky gut?

To understand what people mean by leaky gut, we need to briefly have another look at the gut wall structure that we explained in Chapter 2.

The intestinal (gut) Barrier

The gut barrier is made up of all the components of the gut wall that separate the contents of the gut lumen on one side, from the bloodstream on the other side. Take a quick glance again at Figure 3 in Chapter 1 to remind yourself of the structure of the gut wall. The gut barrier is made up of the single layer of gut epithelium cells on the inside. Sitting on top of these epithelial cells there is a thick layer of mucus (like a layer of sunscreen applied to the skin) that further protects the gut wall. The cells are bound together by lots of complex protein connections, and in particular by structures called 'tight junctions'.

In essence these cells are like tiles, and the tight junctions are the grouting between the tiles.

But this is a simplistic description, because the barrier is not just a passive barrier, it's an active and dynamic structure that has a two-way, highly controlled flow of substances across it, both into and out of the gut.[96] Think of the gut wall as being like a busy toll bridge. The barriers go up and down allowing cars to move in both directions, in a controlled way. The same thing applies to the movement of nutrients, salts, fluid, hormones, and many substances produced by the GM.

Increased gut permeability

If there is inflammation or damage to the gut wall, as we see in Crohn's disease or coeliac disease, this does increase permeability ('leakiness') of the intestinal wall. This is well proven. If there is a lot of inflammation, as we see in patients with Crohn's disease or ulcerative colitis, this also causes significant leakiness of the gut wall and can allow bacterial toxins or even whole bacteria from the gut to cross into the bloodstream. The GM are beneficial when they are in the gut, but can cause serious infection and sepsis if they enter the bloodstream.

Most of the conversations about leaky gut in popular and social media refer to a more subtle increase in gut permeability without any visible damage. This could be caused by damage of the tight junctions (the grouting between the tiles), and it is thought that altered GM (the dysbiosis we discussed previously) might play a role in this. The theory is that an increase in gut permeability allows toxic digestive metabolites, bacterial toxins and proteins and other small molecules to 'leak' into the bloodstream and from there to travel to other organs and parts of the body where they can alter the function of those organs, as well as altering function in the gut. This has been suggested as a possible cause for brain fog, or fibromyalgia, or for some immunological conditions such MS and rheumatoid arthritis, and has also been implicated in autism and a number of other conditions.[97] These areas require a lot of further research before definite conclusions can be made.

Leaky gut and IBS

Gut permeability has been extensively studied in IBS and certainly some people have evidence of increased permeability. If you have IBS or related symptoms and have spent years looking for solutions to your problems, if you read about leaky gut in your search for answers, it will probably sound familiar and resonate with you: gut symptoms, tiredness, brain fog and possible fibromyalgia – suddenly it all makes sense. Leaky gut is the answer...!

The problem with the leaky gut concept is that at this point in time, the balance of evidence suggests that it is more likely that increased gut permeability observed in some people with IBS is the end result of all the processes involved in IBS, rather than the cause.

Is there a treatment for leaky gut?

At this moment in time, there is no proven treatment for leaky gut and we aim to treat the underlying problem rather than treating the leakiness. What's more, there is no routine, reliable test available to diagnose a leaky gut. Many different treatments have been suggested in complementary, alternative and functional medical practices but none of them are proven. The table below shows some of the treatments that have been recommended to treat leaky gut. Remember, there's little or no evidence to support these claims, but this is an evolving area with a huge amount of ongoing research and in ten years' time we may be saying something quite different.

Effect of dietary/other supplements on gut permeability	
Substance	Effect
L-glutamine	Possible minor effect
Psyllium	No proven benefit
Probiotics	Possible minor effect of some *Streptomyces* and lactobacilli
Vitamin c	Possible minor effect
Rhubarb	No proven benefit
Fermented foods (kefir, kimchi & sauerkraut)	No proven benefit

Brain fog: leaky gut, leaky brain?

There is no one precise definition of brain fog or brain fogginess. It's something experienced by many patients with IBS (and in a number of other situations) and describes a number of symptoms that are very poorly understood. Some people describe their brain fog as being very debilitating, whereas for others it is much less of a problem. At the mildest end of the spectrum it includes difficulty concentrating, poor short-term memory and a feeling of thoughts being processed more slowly than normal. At the more severe end of the spectrum patients describe a feeling of mental confusion, poor judgement, a sense of dislocation or almost out-of-body type sensation, even slight slurring of speech and a feeling of unsteadiness on their feet.

Medical professionals have been slow to accept the existence of brain fog or fogginess. Some very interesting research in recent years suggests that brain fog may be related to dysbiosis of the GM and increased permeability of the gut barrier (leaky gut).[98] It's possible that SIBO might also play a role in some people. As we discussed above, increased gut permeability might allow bacterial metabolites, bacterial toxins and other small molecules to pass from the gut into the bloodstream and travel to the brain, where they can alter brain function and lead to a feeling of brain fog. Leaky gut = leaky brain?

Some people with brain fog feel that that their symptoms are brought on by sugars and a high carbohydrate meal, and it has been suggested that fermentation of food products, particularly carbohydrates, by the GM within either the small or large bowel, may lead to the production of substances that are absorbed into the bloodstream and affect brain function.

A recent study raised the possibility of brain fog being triggered by probiotic use. The authors suggested that the probiotics might be colonising the small bowel, causing a SIBO-type effect.[99] This research caused some controversy and hasn't been backed up by other studies, but it is an interesting possibility.[100]

If you have IBS and experience brain fog, it is certainly worthwhile testing for SIBO. Even if the test is negative, if you have persistent symptoms, a trial of treatment for SIBO could still be worthwhile.

A low carbohydrate diet has been recommended for treating brain fog, although there is little hard evidence to support this. Theoretically, by reducing the potential food for bacteria in the small bowel or for the GM, this might reduce the production of 'toxic' substances. The evidence for this is scant though.

Interlinked conditions

There are a number of conditions that affect different systems of the body, but which appear to be strongly interlinked, as they often co-exist with each other; in other words, if you have one of these conditions, you have a higher chance than normal of having another condition. These conditions are all classified as 'functional' and several different umbrella terms have been used, including functional somatic syndrome or disorder or, more recently, they have also been called 'bodily distress syndrome',[101] which strongly evokes the negative effect these conditions can have on someone's overall health and wellbeing. We will just mention two of the more common associated conditions, namely fibromyalgia and chronic fatigue syndrome. All of these conditions are more common in women.

Fibromyalgia

Fibromyalgia, also known as chronic widespread pain (CWP), is characterized by widespread musculoskeletal pain – which includes pains in muscles and cartilage (very often the rib cartilage is involved: a condition called costochondritis). There is usually extreme sensitivity and multiple painful tender areas, called 'trigger-points', which are very painful when examined. Fibromyalgia is more common in women, particularly women with IBS. About half of people with fibromyalgia have IBS and just over half of people with IBS have fibromyalgia.[102] People also often complain of chronic fatigue, poor sleep, low mood and also brain fog (so-called 'fibro-fog'). We spoke earlier about 'visceral hypersensitivity' in patients with IBS; patients with fibromyalgia also have hypersensitivity of the nerves, but this time, the nerves to the muscles, ligaments, joints and skin, and this is called 'somatic hypersensitivity'. As with IBS, dysbiosis has been described in patients who have fibromyalgia.

It is so tough to have unpleasant digestive symptoms from IBS, coupled with fibromyalgia pains and aches all over the body. At this point in time, there is no cure for fibromyalgia, just as there is no cure for IBS, but there are many strategies and treatments available that can help reduce the symptoms.

Chronic fatigue syndrome

Chronic fatigue syndrome (CFS) is described as a constant and persistent fatigue that has been present for more than six months, (we seem very fixated on six months when defining medical conditions!), that is not helped by sleep or rest and for which no underlying medical disorder or nutritional deficiency has been found. Up to 90 per cent of patients with CFS have IBS-type symptoms. A large study, pooling together data from lots of other studies, found that half of patients with IBS also experience fatigue.[103] Fibromyalgia is also common in patients who have CFS. As with IBS and fibromyalgia, dysbiosis has been found in patients with CFS, along with evidence of very low-level inflammation, which is not picked up on standard tests.

If you have fatigue or chronic fatigue, it is extremely important to make sure that you are not deficient of any vitamins – in particular vitamin B12, iron, folate and vitamin D. It is also very important to check for an underactive thyroid. If you mention that you are experiencing fatigue to your GP or family doctor, she or he would generally check this panel of bloods.

Are these symptoms all different manifestations of one condition?

The evidence points strongly towards all of these functional conditions being strongly linked. Increased gut permeability (leaky gut) and altered GM (dysbiosis) may well be playing a role, but it's still not clear if this is the 'chicken or the egg'. There appears to be low-level chronic inflammation in all of the conditions and they are much more common in women than in men. The search goes on to try to find the cause or causes, but in the meantime, we need to acknowledge the substantial loss of wellbeing caused by these conditions, and to try to help people deal with their symptoms as best we can.

Certain medications that act on the somatic, visceral and central nervous systems can be helpful in treating these conditions, particularly certain antidepressant medicines and others that alter pain perception and thresholds. You should discuss this with your doctor.

26

Is it possible to eat too 'cleanly'?
Orthorexia

Many people decide to make major dietary changes for health, environmental or ethical reasons or to control their calorie intake. More and more people are following exclusively plant-based diets, vegetarian diets, high-fibre diets, or low-fat diets for these reasons. These diets can certainly help control weight, but sometimes eating patterns can become excessively restrictive and controlled. Anorexia nervosa has been increasingly diagnosed in recent decades, particularly in young women, but there is another lesser-known condition of disordered eating, called orthorexia nervosa, or more commonly just orthorexia.

Orthorexia

Orthorexia is an unhealthy obsession with eating healthy food also referred to as 'clean eating'. It was first described in 1997 as a 'fixation on eating healthy food in order to avoid ill-health and disease',[104] but has been increasingly recognized since then. It is now thought that as many as one in 100 people may have this condition. It is most common in people working in the health and fitness industry and among dietitians.[105] The name is derived from the Greek word '*orthos*' meaning correct, and, '*orexis*' meaning appetite, suggesting 'correct appetite'.

Orthorexia involves an obsessive focus on healthy eating, defined by a dietary theory or set of beliefs the specific details of which may vary. People sometimes follow 'clean eating' rules with religious fervour, with a fixation on the type, quality and preparation of food to promote what they believe is optimum health. Sometimes people become very stressed or distressed if they cannot follow their own eating rules. Dietary restrictions often escalate over time, and may come to include elimination of entire food groups and involve progressively more frequent fasts to 'detoxify'. Weight loss may occur as a result of adherence to strict dietary guidelines, but is not the primary goal. However, orthorexia can co-exist with other eating disorders such as anorexia, as it may be a way to disguise an underlying eating disorder by adopting what, on the surface, may seem to be more socially acceptable 'healthy' eating patterns.

Over a number of years such restrictive eating patterns can be very detrimental to one's health, causing multiple vitamin deficiencies, with knock-on effects on teeth and bone health, reproductive health, reduced diversity of the GM as well as the effects on mental health. The amount of high-fibre food, fruit and vegetables being eaten by some people can also trigger very significant digestive and IBS-type symptoms, which is why these patients often end up coming to see a doctor or dietitian.

Peer pressure

Orthorexia is more common in people in their 20s/30s than in older people. Social media can play a big role as many different foods are being demonized on various platforms. Perhaps you started by going gluten-free, then decided to become dairy-free as well, then maybe sugar-free, then no or low carbs, then no animal products, and before you know it, all you're eating is large amounts of hard-to-digest plant-based foods, many of which, when consumed in excess, can cause very significant IBS-type symptoms.

Research has found that the healthy-eating community on Instagram has a high prevalence of orthorexia symptoms, thought to be due to its image-based platform.[106] This research found that almost half of Instagram followers of 'healthy eating sites' had features suggestive of orthorexia, compared to only one per cent of other social media platforms. That's an incredibly high percentage.

One 'clean-eating' patient we saw in recent years, told us; 'I was 28 and hadn't had a period in two years, I had osteopenia (thin bones) and my hair was thinning. I was worn down by anxiety about what might happen if I broke my "wellness'" rules'.

This distressed young woman could see what was happening to her body as a result of her eating pattern, and yet felt helpless to stop it.

Becoming more common

We frequently see patients in our clinic who make restrictive dietary choices, very often with good intentions. Orthorexia appears to be increasingly common and excessive 'clean-eating' can result in significant problems with abdominal bloating or diarrhoea. As a result more people with orthorexia are presenting to doctors and dietitians for investigation and treatments of these symptoms.

Lizzie's story

Lizzie, a tall, slim, 24-year-old woman saw Elaine for dietary advice. She was suffering with frequent and loose bowel motions accompanied by abdominal pain and bloating. She was highly concerned by a rapid weight loss of 10kg within a three-month period. Her weight now was 52kg and her BMI 17.3, indicating that she was underweight. She had amenorrhoea (loss of menstrual cycle). Her medical investigations had been normal. Lizzie had just finished college and had started working in a gym for the summer. One of the perks was that she attended a personal trainer for advice in the gym. She embraced a very strict exercise and weights programme and had altered her diet to include higher protein, lower carbohydrates and an increased intake of fruit and vegetables, including smoothies. She told Elaine that all her family were slim and could 'eat what they wanted'.

Elaine explained that some individuals with this type of metabolism can lose weight very quickly when they start exercising regularly. Coupled with reduced carbohydrates and a significantly decreased calorie intake, there is often a rapid weight loss. The gut symptoms were most likely related to too much fibre, poorly absorbed vegetables and a high fructose load.

Elaine discussed making some dietary changes to increase Lizzie's daily calorie intake by 1,000kcal per day, to reintroduce some carbohydrates and to alter the types and quantities of fruit and vegetables. Lizzie became very stressed during the consultation, as she was not happy at the thought of regaining any weight, even though she realized that she was underweight and that her gut symptoms were being triggered by her diet. She acknowledged she was going find the advice difficult to embrace and agreed to attend a psychologist's clinic in conjunction with dietary recommendations and a reduction of her exercise regime.

Lizzie struggled with implementing the recommendations, but made slow and steady progress over six months. Her menstrual cycle returned, she gained 6kg to a weight of 58kg (BMI 19.4) and her gut symptoms improved.

Take-home message Orthorexia can start very easily by embracing a healthy eating and/or exercise plan. It can take hold very quickly, particularly in people who execute advice diligently and have perfectionist traits. It can trigger IBS-type symptoms.

27

Should you eat more fermented or sprouted foods?

Background

Ten years ago, or even five years ago, we would never have considered including a chapter on fermented foods. But there has been such an upsurge in interest that we are asked daily about them by the patients we see in our clinics: 'Should I be eating sauerkraut? I heard that it's really good for gut health', or 'I've been making my own kefir, but it doesn't really seem to be helping my IBS. How long does it take to work?'

The things that we are most commonly asked about, particularly by women with IBS, are fermentation, sprouting and pickling. Let's look at them in a little more detail – but remember there is a difference between something being good for your general health and something being good for your gut symptoms. For example, a fermented food may have been shown to increase antioxidant levels and to be associated with a lower risk of colon cancer, but that same food may cause you abdominal bloating and be distinctly unhelpful if you're struggling to control IBS symptoms.

What is a fermented food?

The International Scientific Association for Probiotics and Prebiotics (ISAPP) recently defined fermented foods (and beverages) as 'foods made through desired microbial growth and enzymatic conversions of food components'.[107] This sounds like a bit of a mouthful, but it simply means that the food is altered by certain microorganisms (either bacteria or fungi) and by the enzymes within them to change the food in a beneficial way. The fact that this is 'desired' microbial activity means that this does not include food simply spoiling or going off (also caused by bacteria).

A brief history of fermented foods

Fermented foods have been part of the human diet for thousands of years, and there are thought to be more than 5,000 varieties of fermented foods currently being produced and consumed by people all over the world. One of the main reasons that fermented foods have been so popular over the centuries is that fermentation is an excellent form of food preservation, and this was very important historically, when food sources were often unreliable. But, equally important to the popularity of fermented foods through history, is the fact that the fermentation process often produces foods that have much more interesting and satisfying tastes and textures than the native food itself. The most common examples of fermented foods are dairy products such as yoghurt, cheese or kefir, but in Eastern cultures there are also many soy-based fermented foods such as tempeh, miso or natto.[108]

Some fermented foods contain live bacteria or fungi, whereas in other fermented foods, the microorganisms, having done their job, are killed during cooking or further processing, and so the fermented food no longer contains any live organisms.

Fermented foods	
Contain live microorganisms	**Live microorganisms absent**
Dairy/milk products • yoghurt, kefir, most cheeses	Bread (sourdough) Wine, most beers and distilled spirits Heat-treated or pasteurized dairy products Heated treated fermented vegetables
Soy products • tempeh, miso, natto	
Vegetables (non heated) • sauerkraut, kimchi	
Beverages • kombucha, some beers	

The fermentation process

Some raw foods naturally contain the microorganisms required to cause fermentation; this is called spontaneous fermentation, or 'wild fermentation'. Examples include sauerkraut, kimchi and some soy products. Other foods are fermented through the addition of starter cultures and this includes most dairy products, sourdough bread, and kombucha. During fermentation, carbohydrate starches or sugars are converted into alcohols or organic acids, and these are responsible for preserving the foods. However, other changes also occur in the foods during fermentation, which have some possible health benefits, and this is one of the reasons that there is such a surge in interest regarding these foods.[109–112]

Why fermented foods are good for health	
Reduce sugar: tackle obesity and Type 2 diabetes	Microorganisms reduce content of high calorie and high-GI sugars in some foods, reducing the GI index.
Lower fructans in bread: less bloating	Sourdough bread has lower fructan levels. This might cause less bloating for people with IBS.
Reduce lactose	Reduction of lactose in dairy products (kefir has 30% lower lactose than milk) may improve tolerability for people with lactose intolerance and IBS.
Lower blood pressure	Some fermented foods have been shown to increase the levels of certain blood pressure-lowering substances.
Live microorganisms	Increased GM diversity.
More vitamins and minerals	Certain bacteria can increase vitamins B12, B9 (folate) & B2 (riboflavin) and vitamin K levels in food. Others can increase the available calcium and magnesium in foods.
More antioxidants	Fermentation by lactobacilli increases polyphenol levels. These powerful antioxidants reduce cancer risk.

Fermented foods and gastrointestinal health
Kefir

What is it? It's a milk or yoghurt drink that has been fermented using kefir grains.

Why might it help? It contains 30 per cent lower lactose levels than ordinary milk and might help those with lactose malabsorption and IBS. It has higher B12 levels than standard milk; another win. It also contains live microorganisms and might improve your GM.

Should I drink it? It can't do any harm even if the benefits aren't clear for IBS. Start with a volume of 100-150mls to see if you tolerate it, without gut symptoms.

Live yoghurts and fermented cheeses

What are these? Fermented live yoghurts and fermented cheeses including most hard and aged cheeses such as traditional cheddars, Gouda, Parmesan and Gruyere contain similar levels of lactose to milk.

Why might they help? Fermented hard cheeses contain very little lactose, and are well tolerated by people with both lactose intolerance and IBS. Fermented live yoghurts actually contain similar lactose levels to milk, but are an excellent source of calcium and have many other health benefits. These also include live microorganisms.

Should I eat them? Yes, please do. These dairy foods have many health benefits and should be included in your diet. But just be aware that cheeses are high in fat and should be eaten in moderation.

Sourdough bread

What is this? Sourdough bread is fermented from natural ingredients and a starter culture that grows over a week.

Why might it help? The prolonged fermentation reduces the fructan content of the bread by up to 90 per cent. It may be easier to digest and reduce IBS symptoms. Just be aware though that this has only been demonstrated in cultures grown for long periods of time, and some commercially available sourdough breads may not have such long culture times and thus may not have substantially lower fructan levels.

Should I eat it? Why not? If you like the taste, give it a try or even try making your own, and see if you tolerate it better than standard bread.

Fermented vegetables

What are these? There are lots of types. Sauerkraut is fermented thinly shredded cabbage. Kimchi is Korean in origin and traditionally contains fermented and salted vegetables such as salted cabbage and chilli.

Why might they help? It's not clear that they help IBS symptoms at all. If sauerkraut and kimchi are heated before eating, they no longer contain any live microorganisms.

Should I eat it? People with IBS often tolerate cabbage poorly, and if you have IBS or significant bloating, we would suggest that you experiment with small amounts of sauerkraut to start with. With kimchi, there's no real information on its effect on IBS, so perhaps try small amounts and see how you feel.

Fermented soy products

What are these? Tempeh, natto and miso are different fermented soy products.

Why might they help? They contain live microorganisms and have been found to have some general health benefits but there is no evidence that they have any specific benefits for gut health or for people with IBS.

Should I eat them? You can certainly give them a try if you enjoy the taste. They are also a good source of protein, particularly helpful if you follow an exclusively plant-based diet.

Sprouting

What is this? Sprouting is a natural process by which seeds germinate and develop in water or another medium and grow sprouts. It has been used in Eastern countries for thousands of years. The process involves soaking grains, legumes, nuts and seeds so that they germinate and 'sprout' little shoots. They are ready to eat when the sprout is 5mm or so.

Why might this help? The sprouting process breaks down some of the indigestible fibres in grains and legumes (in particular the fructans and GOS that we spoke about in previous chapters) meaning that they are more easily digested when we eat them. One exception to this might be chickpeas; sprouting was actually found to increase the GOS content in some cases.[113] The extent of the sprouting affects how much these substances are broken down. Sprouting can also increase the bioavailability of nutrients (that is how much of the nutrients are released from the food to us when we eat them) and have higher concentrations of antioxidants (which have anti-cancer and anti-inflammatory effects).

Should I eat sprouted foods? Going back to first principles it would seem to make sense that the sprouting process might improve tolerance of some foods that are otherwise poorly tolerated by people with IBS, such as the fructan- and GOS- containing foods that we have spoken about previously. Sprouted foods are also rich in many nutrients. So, if you feel passionate about trying sprouted foods you should try them and see if they help *your* tolerance of certain foods.

Pickling

What is this? The pickling process involves putting suitable foods into either brine (a salt solution) or vinegar. If the food is put in brine, then the pickling process involves fermentation by bacteria as well as pickling. If the food is put in vinegar, this usually kills any bacteria/yeast that would cause fermentation. Sauerkraut and kimchi are examples of foods that involve both fermentation and pickling.

Why might it help? Pickling breaks down some of the constituents in foods and may make them easier to digest, but the pickling process can also involve the addition of fructose (which can induce diarrhoea and bloating as we have seen) or acid, which can worsen other digestive symptoms.

Should I eat them? So, a bit like our advice for sprouted foods, try them in small amounts and assess your symptoms. We would not simply embrace these foods because everyone is doing it and tolerability will vary from person to person.

Take-home message

Should you eat fermented, sprouted or pickled foods?

- Some of these foods may be easier to digest.
- Some foods may cause less bloating.
- Some foods (for example sauerkraut) may be less well tolerated, so start with small amounts.
- Effects will vary from person to person, so try a little and see how you feel.
- These foods have other nutritional and health benefits and you should try to include them in your diet if possible.

28

How do I alter my GM? Are probiotics the answer?

Probiotics/prebiotics & faecal microbiota transplantation (FMT)

With all the accumulating evidence of altered GM and dysbiosis in many digestive conditions, but in particular in IBS,[26] it is not surprising that people have been looking at ways of altering the GM and trying to restore it back to a 'normal' or healthy balance. While this is certainly a very admirable goal, we have to realize that the alterations in GM that have been described are varied and complex. It is not simply a matter of too much of one type of bacteria and too little of another, or too much candida.

Probiotics, prebiotics and faecal microbiota transplantation (FMT) are designed to, they claim, alter the GM back to a healthy balance. Can they actually do this? The jury is still out – but it's certainly big business; the probiotic market in the USA is projected to be worth over $67 billion by 2023. The most up-to-date evidence suggests that probiotics don't really 'take root' in the gut, but are more like passengers passing through. This means that if you take a probiotic and find it of some benefit, you will likely need to keep taking it.

Probiotics and prebiotics: the low-down	
Probiotics	**Prebiotics**
What are they?	
Live microorganisms that supposedly have a health benefit when taken in adequate amounts.	Food (mainly different kinds of fibre) or nutritional substances that are used by your GM, resulting in health benefits.

(Continued)

Probiotics and prebiotics: the low-down	
Examples	
Capsules, liquid supplements, live yoghurts: some are single microorganism and some contain multiple strains (multi-strain) of bacteria or fungi.	Various fibre supplements – fructans (inulin) and GOS
How do they work?	
The aim is that the microorganisms in the probiotic reach the colon and exert positive effects there, increasing the healthy GM and interacting in complex ways. Current evidence suggests that they do not take root and mainly exert their effects while you are taking them. Specific effects of probiotics are likely strain-specific and complex; all probiotics are not the same.	By acting as food for the GM, they allow healthy GM, which feed on fibre, to proliferate.
Are there any potential side effects?	
• May be harmful if you have a suppressed immune system • May carry antibiotic resistance genes – a modern day problem, leading to lack of efficacy of many antibiotics • May not reach their target or take root in the bowel (just passing through!) • Expensive • Poorly regulated market: producers make many unfounded claims	Can cause bloating Can worsen IBS symptoms
Do they help IBS?	
Large study suggested that one in seven people with IBS may get some benefit – overall symptoms and abdominal pain may reduce in some people.	We don't know for sure. They might help some people. Effects are only likely to last while taking them.

Probiotics and prebiotics: the low-down	
Are they recommended for IBS?	
The American Gastroenterology Association does not recommend them. The British Society of Gastroenterology says that probiotics might be helpful for abdominal pain and overall IBS symptoms. But they don't work in everyone and *not all probiotic strains are the same. At this point not possible to recommend a particular strain or combination of strains (multi-strain probiotic), but BSG suggests that it's reasonable to try a probiotic for up to 12 weeks to see if there's any effect.*	Not currently as routine.
The Gut Experts' advice in IBS	
We don't recommend them routinely.	We don't recommend them routinely as they can they cause problematic bloating.
Reasonable to take a multi-strain probiotic for one to three months – if no benefit, stop taking it. Could try a different probiotic, but don't do this indefinitely. Do your research. There are so many probiotics on the market and they are not all the same, particularly in potentially helping IBS. If you can (this is where an internet search is helpful) check if there is *any* evidence for the particular probiotic you are thinking of buying, in helping IBS.	We prefer to ensure adequate fibre intake through normal foods and supporting gut health with a healthy, balanced Mediterranean-style diet.

Faecal microbiota transplantation in IBS?

FMT involves taking faeces from a 'healthy' donor and 'transplanting' them along with their healthy GM, to the digestive system of someone with IBS. This is not for the faint hearted! We have carried out FMT for some patients with a nasty bowel infection called *Clostridium difficile*, which has not responded to antibiotics, and there is good data for FMT in that context. The 'donor' stools can be transplanted into the recipient (patient) in a number of ways – a tube via the mouth and stomach into the small bowel, or by spraying it directly into the colon via colonoscopy.

Does it work?

There have been some very good results of FMT for treating IBS symptoms, with some studies showing up to 90 per cent improvement in IBS symptoms. However the longest study followed patients for six months, so at this point, we don't know how long the effects last. Will the symptoms start to come back after a year for example? Is it like getting braces on? Your teeth are perfect after the braces come off, but if you don't use a retainer, they may go back to how they were before. Does the GM in people with IBS treated with FMT gradually go back to how it was before?

It also seems that the faeces donor is important; one of the research studies described their donor as a 'super donor', a healthy young man, who did not smoke, exercised regularly, had a healthy diet and whose GM was found to have a lot of diversity.

A company is currently making 'faeces capsules': small capsules containing stool samples mixed from lots of healthy donors. These are not available commercially at the moment, but this is an area to watch.

Take-home messages

- There is weak evidence to support the use of probiotics in IBS, but only a minority of people will benefit.
- Multi-strain probiotics may be more effective than single strains.
- The effects of probiotics likely to be strain-specific – they are not all the same. Watch this space in the future.
- There is little or no evidence to support use of prebiotics in IBS.
- These treatments can be expensive.
- FMT shows some promise in clinical trials, but duration of effect is not known.

APPENDICES

Appendix 1: Medications used for IBS

There are a myriad of medications that can be used in the treatment of IBS. These should never be used alone for management of IBS, but instead should always be used in combination with the lifestyle and dietary measures that we have just discussed. Many medications are now available without prescription, so-called over-the-counter medications, such as peppermint oil and simple laxatives. Others, such as certain laxatives or anti-spasmodic medications will commonly be prescribed by your family doctor/GP. There are many other medications that are generally prescribed by a specialist, such as a gastroenterologist. The use of medication for treatment of IBS should obviously be under the supervision of a doctor. We will mention a few of the more common medications, rather than an exhaustive list of every possible option. Medications are chosen to help with certain symptoms and this should be done on a person-by-person basis.

Medications used in the treatment of IBS	Examples	Effects
Constipation		
Fibre supplements		
Psyllium/ispaghula husk	Psyllium	Bulking agents
	Fybogel	Hold onto fluid
Iso-osmotic laxatives		
Polyethelene glycol	Movicol	Hold onto fluid in the gut
	Molaxole	
Osmotic laxatives		
Magnesium compounds	Milk of Magnesia	Draw fluid into the gut
Lactulose	Epsom salts	
Stimulant laxative		
Senna	Sennokot	Stimulates gut motility
Bisacodyl	Dulcolax	
Acting on specific gut receptors		

(Continued)

Medications used in the treatment of IBS	Examples	Effects
Lubiprostone	Amitiza	Increases water secretion into gut
Prucalopride	Resolor	Increases motility
Linaclotide	Linzess	Increases water secretion into gut
	Constella	
Diarrhoea		
Antimotility agents		
Loperamide	Imodium	Reduces motility
	Arret	
Opioid-like medications		
Diphenoxylate	Lomotil	Reduces motility
Eluxadoline	Viberzi/Truberzi	
Serotonin receptor blocker (5HT3)		
Alosetron	Lotronex	Reduces motility
Ondansetron	Zofran	
Bile acid sequestrants/'Bile mops'		
Cholestyramine	Questran	'Mops' excess bile in colon
Colesevelam	Cholestagel	
Medications for pain		
Antispasmodics		
Hyoscine	Buscopan	Relaxes gut motility
Mebeverine	Colofac	
Alverine	Spasmonal	
Peppermint oil	Colpermin	Relaxes gut motility
Antidepressant medications		
Tricyclic antidepressants	Amitriptyline	Act on visceral pain receptors
SSRIs	Fluoxetine	and brain–gut axis
	Paroxetine	
	Sertraline	

Choice of medication

The list of medications shown is just a sample of those commonly used in IBS management, and availability may vary from country to country. For example Eluxadoline (Viberzi/Truberzi) is available on prescription in the USA and Canada, but is not approved for use in Europe or Australia. Medications alone are never the answer to managing IBS, but they are a very helpful part of the treatment in some people. The choice of medication obviously depends on the main symptoms.

Constipation

For someone with IBS-C, in addition to looking at dietary fibre and fluid intake, we may recommend using a bulking agent such as Psyllium or a laxative such as polyethylene glycol. These are first-line medications and stronger medications may be needed.

Some people with IBS-C use stimulant laxatives, such as senna or Bisacodyl. These can certainly be effective in stimulating a bowel motion, but this is often associated with cramps. The big drawback of stimulant laxatives is that your body gets used to the medication if it is used regularly and eventually you will need more and more of the medication to have the same effect as one or two tablets had at the outset. This laxative over-use can make the natural motility of the bowel worse over time, leading to a vicious circle of someone needing to take more and more laxatives. We have seen patients who have been taking laxatives for years and who are now taking 40 to 50 laxative pills per day, with no effect! You do not want to get into this situation, and stimulant laxatives should only be used occasionally.

Osmotic laxatives, such as lactulose draw fluid into the bowel and increase the stool volume. Lactulose is metabolized by gut bacteria however, and a drawback of this medication is that it can cause bloating, pain and flatulence.

A number of different laxatives may need to be tried before the right one for you is found.

Diarrhoea

Patients with IBS-D can find their symptoms very debilitating as it can interfere with everyday life to such a large extent. Not to labour the point, but diet is a central part to the management of IBS-D and we have discussed this extensively.

There are a number of medications available and ultimately the choice of medication will depend on your doctor's assessment of the underlying mechanisms. Almost 30 per cent of people who have been diagnosed with IBS-D may have an element of bile salt malabsorption and a trial of 'bile mops' can be very helpful.[114,115]

Simple anti-diarrhoeal medications such as loperamide can be helpful as they slow gut motility. Unfortunately, they do not help the other associated symptoms such as pain and bloating, and in fact can sometimes make these worse. We often recommend using these medications in a preventative way and only from time to time. They can offer reassurance if you are going to be in a position where you don't have ready access to a bathroom and where this is causing you anxiety. We recommended that Mary use loperamide as a preventive medication in this way, in 'Mary's story' in Chapter 9.

Abdominal pain

Abdominal pain may have a number of different causes in someone who has IBS. It might be a cramping or spasm-like pain that occurs before or after a bowel motion. It could also be a more generalized, background pain and discomfort caused by abdominal bloating. For the spasm-like pains, simple anti-spasmodic medications and peppermint oil can be helpful, taken either regularly or when needed. When the pain is more constant or associated with bloating, then a low-dose antidepressant medication can be extremely helpful.

Appendix 2: Medications for functional dyspepsia

There are many different medications available for treatment of functional dyspepsia, and no single medication works in every person, so it can be a question of trial and error. These medications are generally only available on prescription.[82] Your GP or family doctor might prescribe some of these for you as a trial, but if you have more problematic or persistent symptoms, you should be referred to a specialist for appropriate investigation and further treatment as necessary.

Medication	Example	Effects
Simple antacids		
Sodium bicarbonate	Gaviscon	Neutralize gastric acid
Calcium carbonate	Rennies	Can help epigastric pain and burning
Magnesium carbonate	TUMS	
Acid blockers		
Proton pump inhibitors (PPI)	Esomeprazole	Reduce acid production by stomach
	Omeprazole	Can help epigastric pain and burning
	Pantoprazole	
H2 receptor blockers		
	Ranitidine	
	Cimetidine	
	Famotidine	
Anti-nausea		
	Cyclizine	Reduce nausea
	Proclorperazine	
	Ondansetron	
Pro-motility agents		
Dopamine 2 receptor antagonists prandial fullness	Domperidone	Can improve gastric motility & emptying

(Continued)

Medication	Example	Effects
	Metoclopramide	Reduce nausea and post-prandial fullness
Antidepressants (AD)		
Tricyclic AD	Amitriptyline	Reduce epigastric pain and bloating
Tetracyclic AD	Mirtazapine	Improved early satiety, reduced anxiety
Serotonin agonist	Buspirone	Improved early satiety, (5HT1a) nausea and bloating
SSRIs & SNRIs	Paroxetine	No effect on GI symptoms
	Fluoxetine	May help underlying anxiety or mood disorder
	Venlafaxine	

Plant-based therapies (phytotherapy)

There is certainly good evidence to suggest that some natural or plant-based substances have beneficial effects on the symptoms of FD. The mechanism by which these plant-based therapies work is not yet clear, but it is likely through a combination of effects on pain receptors in the stomach, gastric motility and reduction of gas production in the stomach.[116]

Plant-based therapy	Effect
Ginger	Reduces nausea, post-prandial fullness and, possibly, early satiety
Peppermint oil & caraway oil (POCO)	Reduces epigastric pain, fullness, bloating and nausea
Iberogast A mixture of 9 herbs: bitter candytuft, angelica root, caraway, fruit liquorice root, peppermint herb, balm leaf and chamomile flower	Reduces overall symptoms of FD, less pain and fullness
Almonds	Reduces overall FD symptoms

Appendix 3: The FLAT Gut Diet: symptom and lifestyle diary

Step 1 – Simplify

GENERAL RECOMMENDATIONS

- This food, fluid, stress, sleep, exercise and IBS-symptoms diary is designed to provide you with guidance to identify dietary and lifestyle triggers for gut symptoms.
- Keeping this record will help you identify foods that trigger digestive symptoms.
- If you are taking any medications and/or supplements, please record them.

TIPS FOR ACCURATELY FILLING OUT YOUR DIET AND LIFESTYLE DIARY

Guidelines to follow when using this diary:

- Make a note of everything you eat and drink, along with any symptoms.
- Try to fill in your meals, snacks and drinks immediately after eating/drinking them. This way you will not miss anything.
- Make sure to include all drinks.
- Please record the approximate time of eating.
- Fill in as much detail as possible.
- Please state the brand name of all foods e.g. Kellogg's corn flakes, Cadbury's wholenut chocolate.
- Include all foods and snacks even if it is just a handful of crisps or sweets.

INSTRUCTIONS FOR RECORDING *IBS/FD SYMPTOMS* IN YOUR DIARY

When you are recording your food and fluid intake, please also record your symptoms of irritable bowel syndrome and/or FD, at the time they actually occur.

This should be done using the following symbols:

*	normal bowel motion
*c	constipation
**	loose bowel motion
****	very loose bowel motion
A	acid/heartburn
B	bloating
Be	belching
C	cramps
G	gurgling/stomach noises
N	nausea
P	pain
W	wind

Please also indicate severity of symptoms using the following key:

Mild	+
Moderate	++
Severe	+++
e.g. Severe cramps would be	C+++
Mild cramps	C+

Please document *lifestyle* factors:

Alcohol	Standard drinks e.g. 2 glasses wine
Exercise	Time e.g. 20 mins
Stress	0 + ++ +++
Mindfulness/meditation	Time e.g. 5 mins
Sleep	Hours e.g. 6

Step 1 – Simplify: diet, symptom and lifestyle diary

Name: _____ Start date: _____

Week no.: _____

Example of a daily diary

Meals	Monday *Sample Diary*	Symptoms
Breakfast	Tailormade porridge	
	100ml milk	
	10 blueberries	
	5 strawberries	*
Mid-morning	1 banana	
	45g cheddar	
	2 crackers	
Lunch	Poached eggs with roasted vine tomatoes and smashed avocado on 2 slices wholemeal bread	
Mid-afternoon	2 mandarin oranges	
Dinner	Chinese-style chicken with Asian greens	B+
Supper	100g natural yoghurt	
	15g Brazil nuts chopped	
Water (litres)	1.5	
Fibre points	20	
Fructan points	3	
Fructose points	3	
Lactose points	2	
Alcohol	0	
Exercise	30 mins	
Sleep	8 hrs	
Stress rate	++	
Mindfulness	10 mins	

Step 1 – Simplify: diet, symptom and lifestyle diary

Name: _____ Start date: _____

Week no.: _____

Meals		Symptoms
Breakfast		
Mid-morning		
Lunch		
Mid-afternoon		
Dinner		
Supper		
Water (litres)		
Fibre points		
Fructan points		
Fructose points		
Lactose points		
Alcohol		
Exercise		
Sleep		
Stress rate		
Mindfulness		

Meals		Symptoms
Breakfast		
Mid-morning		
Lunch		
Mid-afternoon		
Dinner		
Supper		
Water (litres)		
Fibre points		
Fructan points		
Fructose points		
Lactose points		
Alcohol		
Exercise		
Sleep		
Stress rate		
Mindfulness		

Meals		Symptoms
Breakfast		
Mid-morning		
Lunch		
Mid-afternoon		
Dinner		
Supper		
Water (litres)		
Fibre points		
Fructan points		
Fructose points		
Lactose points		
Alcohol		
Exercise		
Sleep		
Stress rate		
Mindfulness		

Meals		Symptoms
Breakfast		
Mid-morning		
Lunch		
Mid-afternoon		
Dinner		
Supper		
Water (litres)		
Fibre points		
Fructan points		
Fructose points		
Lactose points		
Alcohol		
Exercise		
Sleep		
Stress rate		
Mindfulness		

Meals		Symptoms
Breakfast		
Mid-morning		
Lunch		
Mid-afternoon		
Dinner		
Supper		
Water (litres)		
Fibre points		
Fructan points		
Fructose points		
Lactose points		
Alcohol		
Exercise		
Sleep		
Stress rate		
Mindfulness		

Meals		Symptoms
Breakfast		
Mid-morning		
Lunch		
Mid-afternoon		
Dinner		
Supper		
Water (litres)		
Fibre points		
Fructan points		
Fructose points		
Lactose points		
Alcohol		
Exercise		
Sleep		
Stress rate		
Mindfulness		

Meals		Symptoms
Breakfast		
Mid-morning		
Lunch		
Mid-afternoon		
Dinner		
Supper		
Water (litres)		
Fibre points		
Fructan points		
Fructose points		
Lactose points		
Alcohol		
Exercise		
Sleep		
Stress rate		
Mindfulness		

References

1. Ray K. 'Understanding the immune drivers of food-induced abdominal pain'. *Nat Rev Gastroenterol Hepatol.* 2021;18(3):149. https://doi. org/10.1038/s41575-021-00422-8
2. Anderson S, Cryan J, Dinan T. 'The Psychobiotic Revolution'. *National Geographic*; 2017.
3. Quigley EMM. 'The Gut-Brain Axis and the Microbiome: Clues to Pathophysiology and Opportunities for Novel Management Strategies in Irritable Bowel Syndrome (IBS)'. *J Clin Med.* 2018;7(1):6. https://doi. org/10.3390/jcm7010006
4. Black CJ, Drossman DA, Talley NJ, Ruddy J, Ford AC. 'Functional gastro-intestinal disorders: advances in understanding and management'. *Lancet.* 2020;396(10263):1664–1674. https://doi.org/10.1016/ S0140-6736(20)32115-2
5. Guan Q. 'A Comprehensive Review and Update on the Pathogenesis of Inflammatory Bowel Disease'. *J Immunol Res.* 2019;2019. https://doi. org/10.1155/2019/7247238
6. Mak WY, Zhao M, Ng SC, Burisch J. 'The epidemiology of inflammatory bowel disease: East meets west'. *J Gastroenterol Hepatol.* 2020;35(3):380– 389. https://doi.org/10.1111/jgh.14872
7. Staudacher HM, Whelan K. 'Altered gastrointestinal microbiota in irritable bowel syndrome and its modification by diet: Probiotics, prebiotics and the low FODMAP diet'. In: *Proceedings of the Nutrition Society*. Vol 75. Cambridge University Press; 2016:306–318. https://doi. org/10.1017/S0029665116000021
8. Santos-Marcos JA, Rangel-Zuñiga OA, Jimenez-Lucena R, *et al.* 'Influence of gender and menopausal status on gut microbiota'. *Maturitas.* 2018;116:43–53. https://doi.org/10.1016/j.maturitas.2018.07.008
9. Gill SK, Rossi M, Bajka B, Whelan K. 'Dietary fibre in gastrointestinal health and disease'. *Nat Rev Gastroenterol Hepatol.* 2021;18(2):101–116. https://doi.org/10.1038/s41575-020-00375-4
10. Mills S, Stanton C, Lane JA, Smith GJ, Ross RP. 'Precision nutrition and the microbiome, part I: Current state of the science'. *Nutrients.* 2019;11(4). https://doi.org/10.3390/nu11040923
11. McDonald D, Hyde E, Debelius JW, *et al.* Issue 3 e00031-18 msystems. asm.org 1. msystems. 2018;3. https://doi.org/10.1128/mSystems
12. Caio G, Lungaro L, Segata N, *et al.* 'Effect of gluten-free diet on gut microbiota composition in patients with celiac disease and non-celiac

gluten/wheat sensitivity'. *Nutrients*. 2020;12(6):1–23. https://doi.org/10.3390/nu12061832

13. Sanz Y. 'Effects of a gluten-free diet on gut microbiota and immune function in healthy adult humans'. *Gut Microbes*. 2010;1(3):135–137. https://doi.org/10.4161/gmic.1.3.11868

14. Staudacher HM, Lomer MCE, Anderson JL, *et al.* 'Fermentable carbohydrate restriction reduces luminal bifidobacteria and gastrointestinal symptoms in patients with irritable bowel syndrome'. *J Nutr.* 2012;142(8):1510–1518. https://doi.org/10.3945/jn.112.159285

15. Staudacher HM, Scholz M, Lomer MC, *et al.* 'Gut microbiota associations with diet in irritable bowel syndrome and the effect of low FODMAP diet and probiotics'. *Clin Nutr.* 2021;40(4):1861–1870. https://doi.org/10.1016/j.clnu.2020.10.013

16. Thursby E, Juge N. 'Introduction to the human gut microbiota'. *Biochem J.* 2017;474(11):1823–1836. https://doi.org/10.1042/BCJ20160510

17. Martinez KB, Leone V, Chang EB. 'Western diets, gut dysbiosis, and metabolic diseases: Are they linked?' *Gut Microbes*. 2017;8(2):130–142. https://doi.org/10.1080/19490976.2016.1270811

18. Valdes AM, Walter J, Segal E, Spector TD. 'Role of the gut microbiota in nutrition and health'. *BMJ*. 2018;361:36–44. https://doi.org/10.1136/bmj.k2179

19. Gu Y, Zhou G, Qin X, Huang S, Wang B, Cao H. 'The Potential Role of Gut Mycobiome in Irritable Bowel Syndrome'. *Front Microbiol.* 2019;10(AUG):1894. https://doi.org/10.3389/fmicb.2019.01894

20. Hong G, Li Y, Yang M, *et al.* 'Gut fungal dysbiosis and altered bacterial-fungal interaction in patients with diarrhea-predominant irritable bowel syndrome: An explorative study'. *Neurogastroenterol Motil.* 2020;32(11). https://doi.org/10.1111/nmo.13891

21. Sciavilla P, Strati F, Di Paola M, *et al.* 'Gut microbiota profiles and characterization of cultivable fungal isolates in IBS patients'. *Appl Microbiol Biotechnol.* 2021;105(8):3277–3288. https://doi.org/10.1007/s00253-021-11264-4

22. Auchtung TA, Fofanova TY, Stewart CJ, *et al.* 'Investigating Colonization of the Healthy Adult Gastrointestinal Tract by Fungi'. *mSphere.* 2018;3(2). https://doi.org/10.1128/msphere.00092-18

23. Black CJ, Ford AC. 'Global burden of irritable bowel syndrome: trends, predictions and risk factors'. *Nat Rev Gastroenterol Hepatol.* 2020;17(8):473–486. https://doi.org/10.1038/s41575-020-0286-8

24. Sperber AD, Bangdiwala SI, Drossman DA, *et al.* 'Worldwide Prevalence and Burden of Functional Gastrointestinal Disorders, Results of Rome Foundation Global Study'. *Gastroenterology.* 2021;160(1):99–114.e3. https://doi.org/10.1053/j.gastro.2020.04.014

25. Drossman DA, Morris CB, Schneck S, *et al.* 'International survey of patients with IBS: Symptom features and their severity, health status, treatments, and risk taking to achieve clinical benefit'. *J Clin Gastroenterol.* 2009;43(6):541–550. https://doi.org/10.1097/MCG.0b013e318189a7f9

26. Ford AC, Sperber AD, Corsetti M, Camilleri M. 'Irritable bowel syndrome'. *Lancet.* 2020;396(10263):1675–1688. https://doi.org/10.1016/S0140-6736(20)31548-8

27. Schmulson MJ, Drossman DA. 'What is new in Rome IV'. *J Neurogastroenterol Motil.* 2017;23(2):151–163. https://doi.org/10.5056/jnm16214

28. Ford AC, Lacy BE, Talley NJ. 'Irritable Bowel Syndrome'. Longo DL, ed. *N Engl J Med.* 2017;376(26):2566–2578. https://doi.org/10.1056/NEJMra1607547

29. Zamani M, Alizadeh-Tabari S, Zamani V. 'Systematic review with meta-analysis: the prevalence of anxiety and depression in patients with irritable bowel syndrome'. *Aliment Pharmacol Ther.* 2019;50(2):132–143. https://doi.org/10.1111/apt.15325

30. Koloski NA, Jones M, Talley NJ. 'Evidence that independent gut-to-brain and brain-to-gut pathways operate in the irritable bowel syndrome and functional dyspepsia: a 1-year population-based prospective study'. *Aliment Pharmacol Ther.* 2016;44(6):592–600. https://doi.org/10.1111/apt.13738

31. Black CJ, Yiannakou Y, Guthrie EA, West R, Houghton LA, Ford AC. 'A Novel Method to Classify and Subgroup Patients With IBS Based on Gastrointestinal Symptoms and Psychological Profiles'. *Am J Gastroenterol.* 2021;116(2):372–381. https://doi.org/10.14309/ajg.0000000000000975

32. Black CJ, Yiannakou Y, Guthrie E, West R, Houghton LA, Ford AC. 'Longitudinal follow-up of a novel classification system for irritable bowel syndrome: natural history and prognostic value'. *Aliment Pharmacol Ther.* 2021;53(10):1126–1137. https://doi.org/10.1111/apt.16322

33. Hajizadeh Maleki B, Tartibian B, Mooren FC, *et al.* 'Low-to-moderate intensity aerobic exercise training modulates irritable bowel syndrome through antioxidative and inflammatory mechanisms in women: Results of a randomized controlled trial'. *Cytokine.* 2018;102:18–25. https://doi.org/10.1016/j.cyto.2017.12.016

34. Bennet SMP, Böhn L, Störsrud S, *et al.* 'Multivariate modelling of faecal bacterial profiles of patients with IBS predicts responsiveness to a diet low in FODMAPs'. *Gut.* 2018;67(5):872–881. https://doi.org/10.1136/gutjnl-2016-313128

35. Monsbakken KW, Vandvik PO, Farup PG. 'Perceived food intolerance in subjects with irritable bowel syndrome – Etiology, prevalence and consequences'. *Eur J Clin Nutr.* 2006;60(5):667–672. https://doi.org/10.1038/sj.ejcn.1602367

36. Tuck CJ, Muir JG, Barrett JS, Gibson PR. 'Fermentable oligosaccharides, disaccharides, monosaccharides and polyols: Role in irritable bowel syndrome'. *Expert Rev Gastroenterol Hepatol.* 2014;8(7):819–834. https://doi.org/10.1586/17474124.2014.917956

37. Staudacher HM, Lomer MCE, Farquharson FM, *et al.* 'A Diet Low in FODMAPs Reduces Symptoms in Patients With Irritable Bowel Syndrome and A Probiotic Restores Bifidobacterium Species: A Randomized Controlled Trial'. *Gastroenterology.* 2017;153(4):936–947. https://doi.org/10.1053/j.gastro.2017.06.010

38. Catassi G, Lionetti E, Gatti S, Catassi C. 'The low FODMAP diet: Many question marks for a catchy acronym'. *Nutrients.* 2017;9(3). https://doi.org/10.3390/nu9030292

39. Staudacher HM. 'Nutritional, microbiological and psychosocial implications of the low FODMAP diet'. *J Gastroenterol Hepatol.* 2017;32:16–19. https://doi.org/10.1111/jgh.13688

40. Staudacher HM, Mikocka-Walus A, Ford AC. 'Common mental disorders in irritable bowel syndrome: pathophysiology, management, and considerations for future randomized controlled trials'. *Lancet Gastroenterol Hepatol.* 2021;6(5):401–410. https://doi.org/10.1016/S2468-1253(20)30363-0

41. Tack J. *Why Is Functional Dyspepsia the New Hot Topic? – Rome Foundation.* Published 2020. Accessed May 15, 2021. https://therome-foundation.org/why-is-functional-dyspepsia-the-new-hot-topic/

42. Stanghellini V, Chan FKL, Hasler WL, *et al.* 'Gastroduodenal disorders'. *Gastroenterology.* 2016;150(6):1380–1392. https://doi.org/10.1053/j.gastro.2016.02.011

43. Ford AC, Mahadeva S, Carbone MF, Lacy BE, Talley NJ. 'Functional dyspepsia'. *Lancet.* 2020;396(10263):1689–1702. https://doi.org/10.1016/S0140-6736(20)30469-4

44. Tack J, Schol J, Van den Houte K, Huang IH, Carbone F. 'Paradigm Shift: Functional Dyspepsia – A "Leaky Gut" Disorder?' *Am J Gastroenterol.* 2021;116(2):274–275. https://doi.org/10.14309/ajg.0000000000001077

45. Walker MM, Talley NJ. 'The Role of Duodenal Inflammation in Functional Dyspepsia'. *J Clin Gastroenterol.* 2017;51(1):12–18. https://doi.org/10.1097/MCG.0000000000000740

46. Heitkemper MM, Chang L. 'Do fluctuations in ovarian hormones affect gastrointestinal symptoms in women with irritable bowel syndrome?'

Gend Med. 2009;6(PART 2):152–167. https://doi.org/10.1016/j.genm.2009.03.004

47. Mulak A, Taché Y, Larauche M. 'Sex hormones in the modulation of irritable bowel syndrome'. *World J Gastroenterol.* 2014;20(10):2433–2448. https://doi.org/10.3748/wjg.v20.i10.2433

48. Kim YS, Kim N. 'Sex-gender differences in irritable bowel syndrome'. *J Neurogastroenterol Motil.* 2018;24(4):544–558. https://doi.org/10.5056/jnm18082

49. Baker JM, Al-Nakkash L, Herbst-Kralovetz MM. 'Estrogen–gut microbiome axis: Physiological and clinical implications'. *Maturitas.* 2017;103:45–53. https://doi.org/10.1016/j.maturitas.2017.06.025

50. Ervin SM, Li H, Lim L, *et al.* 'Gut microbial ⊠-glucuronidases reactivate estrogens as components of the estrobolome that reactivate estrogens'. *J Biol Chem.* 2019;294(49):18586–18599. https://doi.org/10.1074/jbc.RA119.010950

51. Derbyshire E, Davies J, Costarelli V, Dettmar P. 'Diet, physical inactivity and the prevalence of constipation throughout and after pregnancy'. *Matern Child Nutr.* 2006;2(3):127–134. https://doi.org/10.1111/j.1740-8709.2006.00061.x

52. Ali RAR, Egan LJ. 'Gastroesophageal reflux disease in pregnancy'. *Best Pract Res Clin Gastroenterol.* 2007;21(5):793–806. https://doi.org/10.1016/j.bpg.2007.05.006

53. Gomes CF, Sousa M, Lourenço I, Martins D, Torres J. 'Gastrointestinal diseases during pregnancy: What does the gastroenterologist need to know?' *Ann Gastroenterol.* 2018;31(4):385–394. https://doi.org/10.20524/aog.2018.0264

54. Houghton LA, Heitkemper M, Crowell MD, *et al.* 'Age, gender, and women's health and the patient'. *Gastroenterology.* 2016;150(6):1332–1343.e4. https://doi.org/10.1053/j.gastro.2016.02.017

55. Landete JM, Arqués J, Medina M, Gaya P, de Las Rivas BD, Muñoz R. 'Bioactivation of Phytoestrogens: Intestinal Bacteria and Health'. *Crit Rev Food Sci Nutr.* 2016;56(11):1826–1843. https://doi.org/10.1080/10408398.2013.789823

56. Setchell KDR. 'The history and basic science development of soy isoflavones'. *Menopause.* 2017;24(12):1338–1350. https://doi.org/10.1097/GME.0000000000001018

57. Sekikawa A, Ihara M, Lopez O, *et al.* 'Effect of S-equol and Soy Isoflavones on Heart and Brain'. *Curr Cardiol Rev.* 2018;15(2):114–135. https://doi.org/10.2174/1573403x15666181205104717

58. El-Salhy M, Gundersen D. 'Diet in irritable bowel syndrome'. *Nutr J.* 2015;14(1). https://doi.org/10.1186/s12937-015-0022-3

59. Moayyedi P, Quigley EMM, Lacy BE, *et al.* 'The effect of fiber supplementation on irritable bowel syndrome: a systematic review and meta-analysis'. *Am J Gastroenterol.* 2014;109(9):1367–1374. https://doi.org/10.1038/ajg.2014.195

60. De Martinis M, Sirufo MM, Viscido A, Ginaldi L. 'Food Allergy Insights: A Changing Landscape'. *Arch Immunol Ther Exp (Warsz).* 2020;68(2). https://doi.org/10.1007/s00005-020-00574-6

61. De Martinis M, Sirufo MM, Suppa M, Ginaldi L. 'New perspectives in food allergy'. *Int J Mol Sci.* 2020;21(4). https://doi.org/10.3390/ijms21041474

62. Lomer MCE. 'Review article: The aetiology, diagnosis, mechanisms and clinical evidence for food intolerance'. *Aliment Pharmacol Ther.* 2015;41(3):262–275. https://doi.org/10.1111/apt.13041

63. Stapel SO, Asero R, Ballmer-Weber BK, *et al.* 'Testing for IgG4 against foods is not recommended as a diagnostic tool: EAACI Task Force Report'. *Allergy Eur J Allergy Clin Immunol.* 2008;63(7):793–796. https://doi.org/10.1111/j.1398-9995.2008.01705.x

64. Atkinson W, Sheldon TA, Shaath N, Whorwell PJ. 'Food elimination based on IgG antibodies in irritable bowel syndrome: A randomized controlled trial'. *Gut.* 2004;53(10):1459–1464. https://doi.org/10.1136/gut.2003.037697

65. Ligaarden SC, Lydersen S, Farup PG. 'IgG and IgG4 antibodies in subjects with irritable bowel syndrome: A case control study in the general population'. *BMC Gastroenterol.* 2012;12. https://doi.org/10.1186/1471-230X-12-166

66. Shepherd SJ, Parker FC, Muir JG, Gibson PR. 'Dietary Triggers of Abdominal Symptoms in Patients With Irritable Bowel Syndrome: Randomized Placebo-Controlled Evidence'. *Clin Gastroenterol Hepatol.* 2008;6(7):765–771. https://doi.org/10.1016/j.cgh.2008.02.058

67. Gibson PR, Shepherd SJ. 'Evidence-based dietary management of functional gastrointestinal symptoms: The FODMAP approach'. *J Gastroenterol Hepatol.* 2010;25(2):252–258. https://doi.org/10.1111/j.1440-1746.2009.06149.x

68. Gibson PR. 'History of the low FODMAP diet'. *J Gastroenterol Hepatol.* 2017;32:5–7. https://doi.org/10.1111/jgh.13685

69. Rej A, Avery A, Ford AC, *et al.* 'Clinical application of dietary therapies in irritable bowel syndrome'. *J Gastrointest Liver Dis.* 2018;27(3):307–316. https://doi.org/10.15403/jgld.2014.1121.273.avy

70. Tuck CJ, Biesiekierski JR, Schmid-Grendelmeier P, Pohl D. 'Food intolerances'. *Nutrients.* 2019;11(7). https://doi.org/10.3390/nu11071684

71. Fassio F, Facioni MS, Guagnini F. 'Lactose maldigestion, malabsorption, and intolerance: a comprehensive review with a focus on current

management and future perspectives'. *Nutrients*. 2018;10(11). https://doi.org/10.3390/nu10111599

72. Fedewa A, Rao SSC. 'Dietary fructose intolerance, fructan intolerance and FODMAPs'. *Curr Gastroenterol Rep*. 2014;16(1). https://doi.org/10.1007/s11894-013-0370-0

73. Cohen SA. 'The clinical consequences of sucrase-isomaltase deficiency'. *Mol Cell Pediatr*. 2016;3(1). https://doi.org/10.1186/s40348-015-0028-0

74. Grant BF, Chou SP, Saha TD, *et al*. 'Prevalence of 12-month alcohol use, high-risk drinking, and DSM-IV alcohol use disorder in the United States, 2001–2002 to 2012–2013: Results from the National Epidemiologic Survey on Alcohol and Related Conditions'. *JAMA Psychiatry*. 2017;74(9):911–923. https://doi.org/10.1001/jamapsychiatry.2017.2161

75. Billings W, Mathur K, Craven HJ, Xu H, Shin A. 'Potential Benefit With Complementary and Alternative Medicine in Irritable Bowel Syndrome: A Systematic Review and Meta-analysis'. *Clin Gastroenterol Hepatol*. Published online September 2020. https://doi.org/10.1016/j.cgh.2020.09.035

76. Wang B, Duan R, Duan L. 'Prevalence of sleep disorder in irritable bowel syndrome: A systematic review with meta-analysis'. *Saudi J Gastroenterol*. 2018;24(3):141–150. https://doi.org/10.4103/sjg.SJG_603_17

77. Lacy BE, Cangemi D, Vazquez-Roque M. 'Management of Chronic Abdominal Distension and Bloating'. *Clin Gastroenterol Hepatol*. 2021;19(2):219–231.e1. https://doi.org/10.1016/j.cgh.2020.03.056

78. Villoria A, Azpiroz F, Burri E, Cisternas D, Soldevilla A, Malagelada JR. 'Abdomino-phrenic dyssynergia in patients with abdominal bloating and distension'. *Am J Gastroenterol*. 2011;106(5):815–819. https://doi.org/10.1038/ajg.2010.408

79. Barba E, Burri E, Accarino A, *et al*. 'Abdominothoracic mechanisms of functional abdominal distension and correction by biofeedback'. *Gastroenterology*. 2015;148(4):732–739. https://doi.org/10.1053/j.gastro.2014.12.006

80. Pesce M, Cargiolli M, Cassarano S, *et al*. 'Diet and functional dyspepsia: Clinical correlates and therapeutic perspectives'. *World J Gastroenterol*. 2020;26(8):456–465. https://doi.org/10.3748/wjg.v26.i5.456

81. Duboc H, Latrache S, Nebunu N, Coffin B. 'The Role of Diet in Functional Dyspepsia Management'. *Front Psychiatry*. 2020;11. https://doi.org/10.3389/fpsyt.2020.00023

82. Ford AC, Luthra P, Tack J, Boeckxstaens GE, Moayyedi P, Talley NJ. 'Efficacy of psychotropic drugs in functional dyspepsia: Systematic review and meta-analysis'. *Gut*. 2017;66(3):411–420. https://doi.org/10.1136/gutjnl-2015-310721

83. Muir JG, Shepherd SJ, Rosella O, Rose M, Barrett JS, Gibson PR. 'Fructan and free fructose content of common Australian vegetables and fruit'. *J Agric Food Chem.* 2007;55(16):6619–6627. https://doi.org/10.1021/jf070623x

84. Loo J Van, Coussement P, De Leenheer L, Hoebreg H, Smits G. 'On the Presence of Inulin and Oligofructose as Natural Ingredients in the Western Diet'. *Crit Rev Food Sci Nutr.* 1995;35(6):525–552. https://doi.org/10.1080/10408399509527714

85. Roberfroid M, Gibson GR, Hoyles L, *et al.* 'Prebiotic effects: Metabolic and health benefits'. *Br J Nutr.* 2010;104(SUPPL.2). https://doi.org/10.1017/S0007114510003363

86. Black CJ, Thakur ER, Houghton LA, Quigley EMM, Moayyedi P, Ford AC. 'Efficacy of psychological therapies for irritable bowel syndrome: Systematic review and network meta-analysis'. *Gut.* 2020;69(8):1441–1451. https://doi.org/10.1136/gutjnl-2020-321191

87. Ford AC, Lacy BE, Harris LA, Quigley EMM, Moayyedi P. 'Effect of Antidepressants and Psychological Therapies in Irritable Bowel Syndrome: An Updated Systematic Review and Meta-Analysis'. *Am J Gastroenterol.* 2019;114(1):21–39. https://doi.org/10.1038/s41395-018-0222-5

88. Manning LP, Yao CK, Biesiekierski JR. 'Therapy of IBS: Is a Low FODMAP Diet the Answer?' *Front Psychiatry.* 2020;11:865. https://doi.org/10.3389/fpsyt.2020.00865

89. Pimentel M, Saad RJ, Long MD, Rao SSC. 'ACG Clinical Guideline: Small Intestinal Bacterial Overgrowth'. *Am J Gastroenterol.* 2020;115(2):165–178. https://doi.org/10.14309/ajg.0000000000000501

90. Rao SSC, Bhagatwala J. 'Small Intestinal Bacterial Overgrowth: Clinical Features and Therapeutic Management'. *Clin Transl Gastroenterol.* 2019;10(10):e00078. https://doi.org/10.14309/ctg.0000000000000078

91. Ghoshal UC, Shukla R, Ghoshal U. 'Small intestinal bacterial overgrowth and irritable bowel syndrome: A bridge between functional organic dichotomy'. *Gut Liver.* 2017;11(2):196–208. https://doi.org/10.5009/gnl16126

92. Takakura W, Pimentel M. 'Small Intestinal Bacterial Overgrowth and Irritable Bowel Syndrome – An Update'. *Front Psychiatry.* 2020;11. https://doi.org/10.3389/fpsyt.2020.00664

93. Giamarellos-Bourboulis EJ, Pyleris E, Barbatzas C, Pistiki A, Pimentel M. 'Small intestinal bacterial overgrowth is associated with irritable bowel syndrome and is independent of proton pump inhibitor usage'. *BMC Gastroenterol.* 2016;16(1). https://doi.org/10.1186/s12876-016-0484-6

94. Ford AC, Harris LA, Lacy BE, Quigley EMM, Moayyedi P. 'Systematic review with meta-analysis: the efficacy of prebiotics, probiotics, synbiotics and antibiotics in irritable bowel syndrome'. *Aliment Pharmacol Ther.* 2018;48(10):1044–1060. https://doi.org/10.1111/apt.15001

95. Chedid V, Dhalla S, Clarke JO, *et al.* 'Herbal Therapy is Equivalent to Rifaximin for the Treatment of Small Intestinal Bacterial Overgrowth'. *Glob Adv Heal Med.* 2014;3(3):16–24. https://doi.org/10.7453/gahmj.2014.019

96. Quigley EMM. 'Leaky gut-concept or clinical entity?' *Curr Opin Gastroenterol.* 2016;32(2):74–79. https://doi.org/10.1097/MOG.0000000000000243

97. Camilleri M. 'Leaky gut: mechanisms, measurement and clinical implications in humans'. *Gut.* 2019;68(8):1516–1526. https://doi.org/10.1136/gutjnl-2019-318427

98. Obrenovich MEM. 'Leaky Gut, Leaky Brain?' *Microorganisms.* 2018;6(4):107. https://doi.org/10.3390/microorganisms6040107

99. Rao SSC, Rehman A, Yu S, De Andino NM. 'Brain fogginess, gas and bloating: A link between SIBO, probiotics and metabolic acidosis article'. In: *Clinical and Translational Gastroenterology.* Vol 9. Nature Publishing Group; 2018. https://doi.org/10.1038/s41424-018-0030-7

100. Quigley EMM, Pot B, Sanders ME. '"Brain Fogginess" and D-Lactic Acidosis: Probiotics Are Not the Cause'. *Clin Transl Gastroenterol.* 2018;9(9). https://doi.org/10.1038/s41424-018-0057-9

101. Budtz-Lilly A, Schröder A, Rask MT, Fink P, Vestergaard M, Rosendal M. 'Bodily distress syndrome: A new diagnosis for functional disorders in primary care? Clinical presentation, diagnosis, and management'. *BMC Fam Pract.* 2015;16(1):1–10. https://doi.org/10.1186/s12875-015-0393-8

102. Hausteiner-Wiehle C, Henningsen P. 'Irritable bowel syndrome: Relations with functional, mental, and somatoform disorders'. *World J Gastroenterol.* 2014;20(20):6024–6030. https://doi.org/10.3748/wjg.v20.i20.6024

103. Whitehead WE, Palsson O, Jones KR. 'Systematic review of the comorbidity of irritable bowel syndrome with other disorders: What are the causes and implications?' *Gastroenterology.* 2002;122(4):1140–1156. https://doi.org/10.1053/gast.2002.32392

104. Bratman S, Knight D. *Health Food Junkies. Orthorexia Nervosa: Overcoming the Obsession with Healthful Eating.* Random House; 2000.

105. Tremelling K, Sandon L, Vega GL, McAdams CJ. 'Orthorexia Nervosa and Eating Disorder Symptoms in Registered Dietician Nutritionists in

the United States'. *J Acad Nutr Diet.* 2017;117(10):1612–1617. https://doi.org/10.1016/j.jand.2017.05.001

106. Turner PG, Lefevre CE. 'Instagram use is linked to increased symptoms of orthorexia nervosa'. *Eat Weight Disord.* 2017;22(2):277–284. https://doi.org/10.1007/s40519-017-0364-2

107. Marco ML, Sanders ME, Gänzle M, *et al.* 'The International Scientific Association for Probiotics and Prebiotics (ISAPP) consensus statement on fermented foods'. *Nat Rev Gastroenterol Hepatol.* 2021;18(3):196–208. https://doi.org/10.1038/s41575-020-00390-5

108. Dimidi E, Cox SR, Rossi M, Whelan K. 'Fermented foods: Definitions and characteristics, impact on the gut microbiota and effects on gastrointestinal health and disease'. *Nutrients.* 2019;11(8). https://doi.org/10.3390/nu11081806

109. Melini F, Melini V, Luziatelli F, Ficca AG, Ruzzi M. 'Health-promoting components in fermented foods: An up-to-date systematic review'. *Nutrients.* 2019;11(5). https://doi.org/10.3390/nu11051189

110. Staudacher HM, Nevin AN. 'Fermented foods: fad or favourable addition to the diet?' *Lancet Gastroenterol Hepatol.* 2019;4(1):19. https://doi.org/10.1016/S2468-1253(18)30392-3

111. Aryana KJ, Olson DW. 'A 100-Year Review: Yogurt and other cultured dairy products'. *J Dairy Sci.* 2017;100(12):9987–10013. https://doi.org/10.3168/jds.2017-12981

112. Marco ML, Heeney D, Binda S, *et al.* 'Health benefits of fermented foods: microbiota and beyond'. *Curr Opin Biotechnol.* 2017;44:94–102. https://doi.org/10.1016/j.copbio.2016.11.010

113. Erba D, Angelino D, Marti A, *et al.* 'Effect of sprouting on nutritional quality of pulses'. *Int J Food Sci Nutr.* 2019;70(1):30–40. https://doi.org/10.1080/09637486.2018.1478393

114. Camilleri M, Vijayvargiya P. 'The Role of Bile Acids in Chronic Diarrhea'. *Am J Gastroenterol.* 2020;115(10):1596–1603. https://doi.org/10.14309/ajg.0000000000000696

115. Lacy BE, Pimentel M, Brenner DM, *et al.* 'ACG Clinical Guideline: Management of Irritable Bowel Syndrome'. In: *The American Journal of Gastroenterology.* Vol 116. NLM (Medline); 2021:17–44. https://doi.org/10.14309/ajg.0000000000001036

116. Kim YS, Kim JW, Ha NY, Kim J, Ryu HS. 'Herbal Therapies in Functional Gastrointestinal Disorders: A Narrative Review and Clinical Implication'. *Front Psychiatry.* 2020;11. https://doi.org/10.3389/fpsyt.2020.00601

Acknowledgements

We would like to thank some very important people who have helped us bring this book to life. Our agent Peter Tallack, who saw the potential of our book and our editor Victoria Roddam at Sheldon Press for her helpful direction and guidance. Dietitian Sam Holman's attention to detail in double and triple checking all the facts and figures has been invaluable.

We have both been very lucky to have great mentors along the way. Elaine would like to thank Dr Heather Holloway and the late Mr Paul O'Byrne who strongly supported her and were great advocates for gut health nutrition. Barbara would like to thank her brother Mark for leading her into medicine, and Prof Nap Keeling for being passionate about IBS, Functional Dyspepsia and the gut-brain axis when no-one else was!

Miriam has been an invaluable part of The Gut Experts' team. We also thank Marese, Maeve and Grace for their unending support in working with us in our clinics. We would like to thank our families for their encouragement and support throughout our careers and over the past two years of writing this book.

And finally – to all our patients, from whom we continue to learn.